PMP® Exam Preparation
Test Questions, Practice Test, and Simulated Exam

PMP® Exam Preparation

Test Questions, Practice Test, and Simulated Exam

Dr. Ginger Levin, PMP, PgMP

CRC Press
Taylor & Francis Group
Boca Raton London New York

CRC Press is an imprint of the
Taylor & Francis Group, an **informa** business

AN AUERBACH BOOK

Parts of *A Guide to the Project Management Body of Knowledge*, 2017, are reprinted with permission of the Project Management Institute, Inc., Four Campus Boulevard, Newtown Square, Pennsylvania 19073-3299 U.S.A., a worldwide organization advancing the state of the art in project management.

"PMBOK" is a trademark of the Project Management Institute, Inc., which is registered in the United States and other nations.

"PMI" is a service and trademark of the Project Management Institute, Inc., which is registered in the United States and other nations.

"PMP" is a certification mark of the Project Management Institute, Inc., which is registered in the United States and other nations.

CRC Press
Taylor & Francis Group
6000 Broken Sound Parkway NW, Suite 300
Boca Raton, FL 33487-2742

© 2019 by Taylor & Francis Group, LLC
CRC Press is an imprint of Taylor & Francis Group, an Informa business

No claim to original U.S. Government works

Printed on acid-free paper

International Standard Book Number-13: 978-0-8153-7910-2 (Paperback)

Visit the Taylor & Francis Web site at
http://www.taylorandfrancis.com

and the CRC Press Web site at
http://www.crcpress.com

Contents

Preface

Based on our experience in helping people to prepare for the Project Management Professional (PMP®) Exam, we know that you will have questions, such as "What topics are covered on the exam?" and "What are the questions like?" Not surprisingly, some of the most sought-after study aids are practice test questions and simulated exams, which are helpful in two ways.

First, taking practice tests increase your knowledge of the kinds of questions, phrases, terminology, and sentence construction that you will encounter on the real exam; and second, taking practice tests provides an opportunity for highly concentrated study by exposing you to a breadth of project management content generally not found in a single reference source.

We developed this specialty publication with one simple goal in mind—that is, to help you study for and pass the PMP® certification exam. Because the Project Management Institute (PMI®) does not sell past exams for prospective certification purposes, the best option is to develop practice test questions that are as representative as possible. And that is exactly what we have done.

This guide contains 40 practice multiple-choice questions for each of the ten knowledge areas in PMI's *A Guide to the Project Management Body of Knowledge (PMBOK®) Guide Sixth Edition,* We have also included a 200-question representative practice test that is in the book in addition to simulated exams based on this practice exam. The simulated exams are four hours, and a clock shows the amount of time you have remaining to complete the exam. At the end, you will see your score by process group and also according to our proprietary algorithm. Our algorithm shows whether you were Above Target, Target, Below Target, or Needs Improvement in each of the five process groups: Initiating, Planning, Executing, Monitoring and Controlling, and Closing.

As we have done in our other PMP® Exam Practice Test & Study Guide books, we have included a plainly written rationale for each correct answer along with a supporting reference list. Each answer also contains a reference to the PMI® *PMP Examination Content Outline,* 2015. Other references are provided at the end of this book.

Earning the PMP® certification is a prestigious accomplishment but studying for it should not be difficult if you use the tools available.

Good luck on the exam! We hope you are the next PMP.

Dr. Ginger Levin, PMP, PgMP
Lighthouse Point, Florida

Acknowledgements

I want to acknowledge the efforts of our publisher, CRC Press, and especially that of Richard Hanley, Randy Burling, and Stephanie Place-Retzlaff, and the entire CRC team who worked tirelessly to publish this book, so it would be available based on the Sixth Edition of the PMBOK® Guide.

I also want to thank my long-time and great friends, Gary and Judy Rechtfertig, who volunteered their time to proofread my work and double check the page number references; their help was invaluable.

Finally, I want to thank my long-time friend, J. LeRoy Ward, who is now with another company, but had the idea of publishing a PMP Study Guide book organized by knowledge areas after the 1996 PMBOK Guide was released.

Online Practice Test

To access this practice test, please contact pmpsimulatedexam@ittoday.info. The online test is four hours and has a clock that shows how much time you have left in the four hours. Once you have answered all of the questions or the time ends, you will be scored by domain and also through our proprietary algorithm that assesses your strengths and weaknesses to identify the areas in which you require further study.

About the Author

Dr. Ginger Levin is an author, senior consultant, and educator in portfolio, program and project management with more than 50 years' experience in the public and private sectors. Her specialty areas include program management, business development, maturity assessments, metrics, organizational change, knowledge management, and the project management office. She is an Adjunct Professor for the University of Wisconsin-Platteville in its master's degree program in project management. Before her consulting and teaching career, she was President of GLH, Incorporated, a woman-owned small business in the Washington, DC area for 15 years, specializing in project management. Earlier, she had a career in the U.S. Government, working for six agencies in positions of increasing responsibility for 14 years. Dr. Levin is the author, co-author, or editor of more than 20 books in the project profession.

Dr. Levin is a member of and active volunteer for the Project Management Institute (PMI®) and a frequent speaker at PMI® Congresses, Chapters, and on projectmanagement.com. In 2014, she was the proud recipient of PMI' Eric Jennett Award.

She is certified by PMI® as a Project Management Professional (PMP®), and a Program Management Professional (PgMP®), and she was the second person in the world to earn the PgMP® designation.

Dr. Levin holds a Doctorate in Public Administration from The George Washington University, where she also received the Outstanding Dissertation Award from the School of Government and Business Administration. She also holds a Master of Science in business administration degree from The George Washington University and a Bachelor of Business Administration degree from Wake Forest University.

Acronyms

AC	actual cost
AD	activity duration
BAC	budget at completion
CEO	chief executive officer
CPI	cost performance index
CPM	critical path method
CV	cost variance
EAC	estimate at completion
ECO	Examination Content Outline
EMV	expected monetary value
ERP	enterprise resource planning
ETC	estimate to complete
EV	earned value
EVA	economic value added
EVM	earned value management
IFD	invitation for bid
OBS	organizational breakdown structure
PDM	precedence diagramming method
PgMP®	Program Management Professional
PMBOK® Guide	*A Guide to the Project Management Body of Knowledge*
PMI®	Project Management Institute
PMIS	project management information system
PMO	project management office
PMP®	Project Management Professional
PV	planned value
RACI	responsible, accountable, consult, inform
RAM	responsibility assignment matrix
RBS	resource breakdown structure
RBS	risk breakdown structure
RFI	Request for information
RFP	Request for proposal
RFQ	Request for quotation
ROI	return on investment

SPI	schedule performance index
SV	schedule variance
SWOT	strengths-weaknesses-opportunities-threats
TCPI	to-complete performance index
TOR	terms of reference
VAC	variance at completion
WBS	work breakdown structure

Introduction

The PMP® exam contains 200 questions, of which 25 questions will not be included in the pass/fail determination. These "pretest" items, as PMI® calls them, will be randomly placed throughout the exam to gather statistical information on their performance to determine their use for future exams. The questions on the exam are distributed following process group in the *PMI PMP Examination Content Outline (ECO)—April 2015.*

- 13% or 26 questions relate to Initiating the Project
- 24% or 48 questions relate to Planning the Project
- 31% or 62 questions relate to Executing the Project
- 25% or 50 questions relate to Monitoring and Controlling the Project
- 7% or 14 questions relate to Closing the Project

For the practice test in this book, we provide all 200 questions as if they were real questions, and the percentages above are applied to the 200 questions. There are no pretest questions in our practice exam. We also have linked each question to a specific task in one of the domains in the ECO especially since some of the questions on the PMP exam are based on the tasks in the ECO, and some tasks are not covered in the *A Guide to the Project Management Body of Knowledge (PMBOK® Guide)* Sixth Edition.

There also are not a certain number of the scored 175 questions that you must answer correctly. PMI® explains in its *PMP Credential Handbook.* It a score based on proficiency levels in the answers in each of the five process groups or domains. This approach means that you will learn whether you were Above Target, Target, Below Target, or Needs Improvement in each process group.

We do not know the algorithm PMI® uses to establish the proficiency levels. Our simulated exam, therefore, scores each question equally, which means you will learn the number of questions you answer correctly in each of the five process groups. Obviously, your goal is to be Above Target, and it also is our goal. In our on-line simulated exam, we have designed our own proprietary algorithm, so you will see your score by domain in the four areas.

To use the study guide effectively, work on one section at a time. It does not matter which one you choose first. Then answer the 40 practice questions by knowledge area, recording your answers on the sheet provided. Finally, compare your answers with those in the answer key. The rationales provided should clarify any misconceptions you may have had, and the process group or domain designations will give you an understanding of the types of questions you might see on the exam that relate to those process groups.

After you have finished answering the questions that follow each section, it is time to take the completely rewritten on-line test and the 200-question practice test in the book that has the number of questions by domain or process group in the ECO. Note your answers on the sheet provided, compare your answers to the answer key, and use the Study Matrix in the Appendix to determine what areas you need to study further.

To access the on-line test please contact pmpsimulatedexam@ittoday.info. The on-line test has a clock that ticks down and shows you how much time you have left in the four hours for the exam. You can take the on-line test as often as you wish and see the questions in a different order.

Our suggestion is to score at least 80% of the questions in our practice test and on-line test correctly before you take the PMP® exam.

We also suggest that when you go to the testing center, you go with a completely positive attitude. By this time, you have done your very best to master the project management concepts you will see on the exam.

Project Integration Management

Practice Questions

INSTRUCTIONS: Note the most suitable answer for each multiple-choice question in the appropriate space on the answer sheet.

1. You work for a software development company that has followed the waterfall development model for more than 20 years. Lately, a number of customers have complained that your company is taking too long to complete its projects. You attended a class on agile development methods and believe that if the company used the agile approach, it could provide products to clients in a shorter time period. However, it would be a major culture change to switch from the waterfall methodology to the agile approach and to train staff members in this new approach. You mentioned this idea to the director of the PMO, and although she liked the idea, she would need approval from the company's portfolio review board to move forward with it. She suggested that you document this idea in a—

   a. Business need description
   b. Product scope description
   c. Project charter
   d. Business case

2. You are managing a large project with 20 key internal stakeholders, eight contractors, and six team leaders. You must devote attention to effective integrated change control. This means you are concerned primarily with—

   a. Reviewing, approving, and managing changes
   b. Maintaining baseline integrity, integrating product and project scope, and coordinating change across knowledge areas
   c. Integrating deliverables from different functional specialties on the project
   d. Establishing a change control board that oversees the overall project changes

1

3. You plan to hold a series of meetings as you execute the project plan. While different attendees will attend each meeting, a best practice to follow is to:

   a. Group stakeholders into categories to determine which ones should attend each meeting
   b. Ensure the attendees have a defined role
   c. Be sensitive to the fact that stakeholders often have very different objectives and invite them to determine the meeting's agenda
   d. Recognize that roles and responsibilities may overlap so focus on holding meetings primarily for decision making

4. You are the project manager in charge of developing a new shipping container for Globus Ocean Transport, which needs to withstand winds of 90 knots and swells of 30 meters. In determining the dimension of the container and the materials to be used in its fabrication, you convene a group of knowledgeable professionals to gather initial requirements, which will be included in the—

   a. Project charter
   b. Bill of materials
   c. WBS
   d. Project Statement of Work

5. You have assembled a core team to develop the project management plan for the next generation of fatigue fighting drugs. The science is complex, and the extended team of researchers, clinicians, and patients for trials exceeds 500 people. In your plan you will be including a number of components, one of which is the life cycle. Of the following which is the most appropriate life cycle to select

   a. Predictive
   b. Iterative
   c. Incremental
   d. Hybrid

6. When you established the change control board for your avionics project, you established specific procedures to govern its operation. The procedures require all approved changes to baselines to be reflected in the—

   a. Performance measurement baseline
   b. Change management plan
   c. Quality assurance plan
   d. Project management plan

7. You are beginning a new project staffed with a virtual team located across five countries. To help avoid conflict in work priorities among your team members and their functional managers, you ask the project sponsor to issue a—

    a. Memo to team members informing them that they work for you now
    b. Project charter
    c. Memo to the functional managers informing them that you have authority to direct their employees
    d. Human resource management plan

8. The benefit of Manage Project Knowledge is to—

    a. Leverage prior organizational knowledge
    b. Prepare a lessons-learned register
    c. Create new knowledge to achieve project objectives
    d. Contribute to organizational learning

9. Interpersonal and team skills are used throughout project management. Your company is embarking on a project to completely eliminate defects in its products. You are the project manager for this project, and you are developing your project charter. To assist you, which of the following interpersonal and team skills do you plan to use—

    a. Surveys
    b. Brainstorming
    c. Meeting management
    d. Focus groups

10. The Direct and Manage Project Work process truly is important in project management. It affects many other key processes and uses inputs from others. Working with your team at its kickoff meeting, you explain the key benefit of this process is to—

    a. Implement approved changes
    b. Provide overall management of the project work
    c. Lead and perform activities in the project management plan
    d. Perform activities to accomplish project objectives

11. You are managing a project in an organization characterized by rigid rules and policies and strict supervisory controls. Your project, sponsored by your CEO who is new to the company, is to make the organization less bureaucratic and more participative. You are developing your project management plan. Given the organization as it now is set up, as you prepare your plan, you can use which of the following organizational process assets—

    a. Guidelines and criteria
    b. Project management body of knowledge for your industry
    c. Organizational structure and culture
    d. The existing infrastructure

12. You are fairly new to managing a project but have been a team member for many years. You are pleased you were selected to manage your company's 2022 model line of hybrid vehicles. You are now planning your project and have been preparing the subsidiary plans as well. You realize some project documents also are required to help manage your project. An example of one that you believe will be especial helpful is the—

    a. Business case
    b. Key performance indicators
    c. Project management information system
    d. Assumptions log

13. You work for a telecommunications company, and when developing a project management plan for a new project, you found that you must tailor some company processes because the product is so different than those products typically produced by your company. Even so, you will consider as you prepare your project management plan—

    a. Standard policies, processes, and procedures
    b. Stakeholder risk tolerances
    c. Expert judgment
    d. Structure of your company

14. You are implementing a project management methodology for your company that requires you to establish a change control board. Which one of the following statements best describes a change control board?

    a. Recommended for use on all (large and small) projects
    b. Used to review, approve, or defer change requests
    c. Managed by the project manager, who also serves as its secretary
    d. Reviews configuration management activity

15. An automated tool, project records, performance indicators, data bases, and financials are examples of items in—

    a. Organizational process assets
    b. Project management information systems
    c. Project management planning approaches
    d. The tools and techniques for project plan development

16. You realize that projects represent change, and on your projects, you always seem to have a number of change requests to consider. In your current project to manage the safety of the nation's cheese products and the testing methods used, you decided to prepare a formal change management plan. An often-overlooked type of change request is—

    a. Adding new subject matter experts to your team
    b. Updates
    c. Work performance information
    d. Enhancing the reviews performed by your project's governance board

17. You have been directed to establish a change control system for your company, but you must convince your colleagues to use it. To be effective, the change control system must include—

    a. Procedures that define how project documents may be changed
    b. Specific change requests expected on the project and plans to respond to each one
    c. Performance reports that forecast project changes
    d. A description of the functional and physical characteristics of an item or system

18. You are working on the next generation of software for mobile phones for your telecommunications company. While time to market is critical, you know from your work on other projects that management reviews can be helpful and plan to use them on your project. You are documenting them as part of your—

    a. Governance plan
    b. Change management plan
    c. Performance reviews
    d. Project management plan

19. Your cost control specialist has developed a budget plan using cost / benefit analysis for your project to add a second surgical center to the Children's Hospital. As you analyze cash flow requirements, you notice that cash flow activity is high. Your business case, which was approved, showed the need for financial resources would be great so you decided to prepare a benefit realization plan for your executives, team, and stakeholders. The purpose of this plan in conjunction with the project management plan is to—

    a. Provide the basis to measure success
    b. Show key roles and responsibilities
    c. Address options for opportunities
    d. Describe the business value resulting from the project

20. You are project manager for a systems integration effort and need to procure the hardware and software components from external sources. Your project is complex as you will have over 20 contractors involved. Your project team is experienced, and you will be working to continually demonstrate the project's benefits to the management team. You have found many do not read the detailed plans you prepare so you decided to—

    a. Use visual management tools
    b. Prepare a balanced scorecard
    c. Set key performance indictors in your charter and provide weekly reports to focus on your progress
    d. Enable greater access to your project management information system

21. Because your project is slated to last five years, you believe rolling wave planning is appropriate. It provides information about the work to be done—

    a. Throughout all project phases
    b. For successful completion of the current project phase
    c. For successful completion of the current and subsequent project phases
    d. In the next project phase

22. Variance analysis is used throughout project management and is a useful tool and technique. Assume you decide to use it as you make decisions as to how to best monitor and control the work on your project. It is helpful because—

    a. It can serve as a basis to reject requested changes
    b. It can identify the reasons for a deviation
    c. It helps determine the type of corrective actions to use
    d. It provides an integrated perspective

23. You are managing a project to introduce a new product to the marketplace that is expected to have a very long life. In this situation, the concept of being *temporary,* which is part of the definition of a project,—

    a. Does not apply because the project will have a lasting result
    b. Does not apply to the product to be created
    c. Recognizes that the project team will outlive the actual project
    d. Does not apply because the project will not be short in duration

24. When closing a project, it is a best practice to—

    a. Update the project documents
    b. Prepare a sustainment plan for the project's benefits
    c. Measure product scope against the project management plan
    d. Review the project management plan

25. Many components of the project management plan are outputs of other processes. However, some other components are equally important and are prepared as part of the Develop Project Management Plan process. An example is the—

    a. Performance management baseline
    b. Scope baseline
    c. Resource breakdown structure
    d. Lessons learned register

26. You are responsible for a project management training curriculum that is offered throughout the organization. In this situation, your intangible deliverables are—

    a. Employees who can apply the training effectively
    b. Training materials for each course
    c. Certificates of completion for everyone who completes the program
    d. The training curriculum as advertised in your catalog

27. Working on your project management training curricula project, you decided it would be beneficial to you to become an active member of the Project Management Institute as part of the objectives of your project is to ensure it is aligned with PMI®'s best practices. To complement PMI®'s *Work Breakdown Structure Practice Standard,* you learned PMI® was requesting volunteers to participate in development of a similar standard on the Scope Statement. You volunteered, and now the Standard is issued. This is an example of:

    a. Improving your own competency as a project manager
    b. Corrective action
    c. Preventive action
    d. A requirement for you to immediately update your project management plan

28. Ideally, a project manager should be selected and assigned at which point in the project life cycle?

    a. During the initiating processes
    b. During the project planning process
    c. At the end of the concept phase of the project life cycle
    d. Prior to the beginning of the development phase of the project life cycle

29. Assume your seven-year project to build the next generation of airplanes that can fly anywhere in the world for 24 hours is complete. You now are closing the program. You have decided to use which of the following tools and techniques to assist you in this final process—

    a. Regression analysis
    b. Document updates
    c. Benefit management plan
    d. Issue log

30. As you are working on your telecommunications project, even though you are using agile methods, you realize you are preparing an extensive amount of data and information. You regularly share data with your project team. Your last team meeting focused on the number of change requests and also the start and finish dates of activities in your schedule. They are examples of—

    a. Key performance indicators
    b. Work performance reports
    c. Work performance data
    d. Work performance information

31. Assume your program management plan has been approved, and you are now managing the work to be done to complete the project successfully. You realize your work is affected by the area in which you work, which is in pork safety and is affected by possible new regulations and tariffs in your country, which may limit your possible exports. You therefore are using your—

    a. Requirements traceability matrix
    b. Project charter
    c. Resource breakdown structure
    d. Geographic distribution of facilities

32. Assume your project to redesign your headquarters so it has state-of-the-art technology is complete, and you are in the closing phase. You decided to hold some meetings in order to—

    a. Confirm the deliverables have been accepted
    b. Transition your work to ongoing operations in your organization for maintenance
    c. Update project documents
    d. Prepare the final report

33. You are a personnel management specialist recently assigned to a project team working on a team-based reward and recognition system. The other team members also work in the human resources department. The project charter should be issued by—

    a. The project manager
    b. The client
    c. A sponsor
    d. A member of the PMO who has jurisdiction over human resources

34. Your project is proceeding according to schedule. You have just learned that a new regulatory requirement will cause a change in one of the project's performance specifications. This change request has been approved. You have had a number of change requests, such as this one, plus issues to handle. A key difference between the two is that—

    a. Issues involve all risks, while change requests are used for many actions
    b. Issues are found when work is being performed, but change requests can be submitted to modify documents, deliverables, or baselines
    c. Change requests and the actions taken to handle them become part of the issue log
    d. Issues are unexpected and handled by the project manager, while change requests are handled by a configuration management board

35. Different types of project processes are used on projects. Some processes are used once at predetermined times, some are performed periodically on the project, and others are performed continuously. An example of the latter is—

    a. Acquire resources
    b. The project charter
    c. Conduct procurements
    d. Monitoring and controlling

36. Assume your company is a leader in the market in production of cereal products. It has been in this market for over 50 years. You are the project manager for a new product that is a derivative from the company's core product. As you determine a life cycle for this project, you believe you should follow one that is—

    a. Incremental
    b. Predictive
    c. Iterative
    d. Adaptive

37. Assume your project consisted of producing a new generation of cell phones that could access the internet from anyplace at any time. However, you learned today that your company acquired a competitor, and the competitor has a comparable product that is being manufactured. You were then told to terminate your project. In this situation you should—

    a. Notify all stakeholders
    b. Update the closure documents
    c. Conduct an immediate review of the work packages
    d. Provide performance evaluations of your team even though you worked in a weak matrix

38. On your project you want to avoid bureaucracy, so you adopt an informal approach to change control. The main problem with this approach is—

    a. There is no "paper trail" of change activity
    b. Regular disagreements between the project manager and the functional manager will occur
    c. There are misunderstandings regarding what was agreed upon by stakeholders
    d. There is a lack of sound cost estimating to assess the change's impact

39. Your project to establish a knowledge transfer process and a knowledge repository is complete. You performed a number of pilot tests, and people in your company are pleased with the results. You are receiving a promotion to become the company's first Chief Knowledge Officer once the project officially is closed. In doing so, you are following the company's activities for administrative closure, one of which is—

    a. Measuring customer satisfaction
    b. Measuring the project's benefits
    c. Evaluating the change requests
    d. Conducting quality control measurements

40. All projects involve some extent of change because they involve work that is unique in some fashion. Therefore, it is important that a project management plan includes a—

    a. Description of the change request process
    b. Configuration management plan
    c. Methodology for preventive action to avoid the need for excessive changes
    d. A work authorization system

Answer Sheet

1.	a	b	c	d
2.	a	b	c	d
3.	a	b	c	d
4.	a	b	c	d
5.	a	b	c	d
6.	a	b	c	d
7.	a	b	c	d
8.	a	b	c	d
9.	a	b	c	d
10.	a	b	c	d
11.	a	b	c	d
12.	a	b	c	d
13.	a	b	c	d
14.	a	b	c	d
15.	a	b	c	d
16.	a	b	c	d
17.	a	b	c	d
18.	a	b	c	d
19.	a	b	c	d
20.	a	b	c	d

21.	a	b	c	d
22.	a	b	c	d
23.	a	b	c	d
24.	a	b	c	d
25.	a	b	c	d
26.	a	b	c	d
27.	a	b	c	d
28.	a	b	c	d
29.	a	b	c	d
30.	a	b	c	d
31.	a	b	c	d
32.	a	b	c	d
33.	a	b	c	d
34.	a	b	c	d
35.	a	b	c	d
36.	a	b	c	d
37.	a	b	c	d
38.	a	b	c	d
39.	a	b	c	d
40.	a	b	c	d

Answer Key

1. d. Business case

   The business case is used to provide the necessary information to determine whether a project is worth its investment. It is used to justify the project and typically contains business needs, analysis of the situation, identifying options for consideration about the business problem or opportunity, an evaluation, and a recommendation, which describes the plan to measure the benefits the project will deliver. [Initiating]

   PMI®, *PMBOK® Guide*, 2017, 30–32
   PMI® *PMP Examination Content Outline*, 2015, Initiating, 5, Task 1

2. a. Reviewing, approving, and managing changes

   Performing integrated change control consists of coordinating and managing changes across the project and communicating the decisions that are made. It involves managing changes to deliverables, documents, and the project management plan. Changes are considered in an integrated way and address risks that may arise from the change. It is performed throughout the project, and the change requests may impact product scope and project scope. Any stakeholder may request a change, and the extent of change control is determined by the project's complexity, contract requirements, and the environment in which the work is done. Once baselines are set, changes follow this formal process. [Monitoring and Controlling]

   PMI®, *PMBOK® Guide*, 2017, 111–113
   PMI® *PMP Examination Content Outline*, 2015, Monitoring and Controlling, 9, Task 1

3. b. Ensure the attendees have a defined role

   Meetings are a tool and technique used in Direct and Manage Project Work. Meetings may include kick-offs, technical, sprint or iteration, Scrum daily standups, steering group, decision making, project updates, or retrospective ones. If a meeting has too many participants, it will not be effective. Each attendee then should have an active role to be invited to each meeting. [Executing]

   PMI®, *PMBOK® Guide*, 2017, 95
   PMI® *PMP Examination Content Outline*, 2015, Executing, 8, Task 6

4. a. Project charter

   The project charter documents the project purpose; measurable objectives/success criteria; high-level requirements; overall risk; summary milestone schedule; preapproved financial resources; project exit criteria; assigned project manager, responsibility, and authority level; and the name and authority of the person who authorizes the charter. It is the document used to formally authorize the project. [Initiating]

   PMI®, *PMBOK® Guide*, 2017, 81
   PMI® *PMP Examination Content Outline*, 2015, Initiating, 5, Task 5

5. d. Hybrid

   Life cycles are series of phases projects go through from the start to completion. This situation with the complexities and the 500-people involved make the hybrid life cycle the preferred style to use. In a hybrid life cycle, it combines predictive and adaptive. With predictive in this case, scope, time, and cost are determined early, and in adaptive with the defined scope are ones that are agile or change driven. With over 500 people involved, changes will be numerous, and, in this situation, regulatory approval also is required. Other life cycles are iterative and incremental. [Planning]

   PMI®, *PMBOK® Guide*, 2017, 74, 88
   PMI® *PMP Examination Content Outline*, 2015, Planning, 7, Task 11

6. d. Project management plan

   All baselines are part of the project management plan. During monitoring and controlling, if there are changes because of the need to take corrective or preventive action, they are reflected in the project management plan. A change request is prepared, and it follows the change control process to update the project management plan. Project documents that are updated are the cost forecasts, issue log, lessons learned register, risk register, and schedule forecasts. [Monitoring and Controlling]

   PMI®, *PMBOK® Guide*, 2017, 97, 112–113
   PMI® *PMP Examination Content Outline*, 2015, Monitoring and Controlling, 9, Task 2

7. b. Project charter

Although the project charter cannot stop conflicts from arising, it can provide a framework to help resolve them, because it describes the project manager's authority to apply organizational resources to project activities. In addition, it documents the purpose of the project; project objectives; high-level requirements; high-level description of the project; overall project risks; summary milestones; preapproved financial resources; key stakeholder list; project approval requirements; project exit criteria; and the name and authority of the sponsor. [Initiating]

PMI®, *PMBOK® Guide*, 2017, 81
PMI® *PMP Examination Content Outline*, 2015, Initiating, 5, Task 5

8. a. Leverage prior organizational knowledge

In the Manage Project Knowledge process, its benefits are to leverage prior organizational knowledge to produce or improve the project outcomes, and to ensure knowledge created by the project is used to support organizational operations and future projects or phases. [Executing]

PMI®, *PMBOK® Guide*, 2017, 98
PMI® *PMP Examination Content Outline*, 2015, Executing, 8, Task 8

9. c. Meeting management

Meeting management is an example of interpersonal and team skills used in developing the project charter. It involves preparing an agenda, ensuring a representative from each stakeholder group is invited, and preparing and sending follow-up notes and action items. Other interpersonal and team skills used are conflict management and facilitation. [Initiating]

PMI®, *PMBOK® Guide*, 2017, 80
PMI® *PMP Examination Content Outline*, 2015, Initiating, 5, Task 5

10. b. Provide overall management of the project work

While all of the answers apply to the Direct and Manage Project Work process, the key benefit is that it involves providing overall management of the work of the project and deliverables to improve the probability of project success. In this question, the key words were "the key benefit". Had the question instead asked about the purpose of the project, answer c would have been the correct choice. [Executing]

PMI®, *PMBOK® Guide*, 2017, 90
PMI® *PMP Examination Content Outline*, 2015, Executing, 8, Task 2

11. a. Guidelines and criteria

While you are managing a different type of project, the organization has managed projects before and therefore may have as part of its organizational process assets a project management template, which sets forth guidelines and criteria to tailor the organization's processes to satisfy specific needs of the project. In addition, project closure guidance or requirements, such as the product validation and acceptance criteria, may be part of organizational process assets for use in the template for the project management plan. [Planning]

PMI®, *PMBOK® Guide*, 2017, 84
PMI® *PMP Examination Content Outline*, 2015, Planning, 7, Task 11

12. d. Assumptions log

The assumptions log is used to record all assumptions and constraints in the project life cycle. Assumptions are factors in the planning process considered to be true, real, or certain even if proof or demonstration is not available. This log is an output from the Develop Project Charter process. It includes high-level and operational assumptions in the business case as well as lower-level and task assumptions that are generated during the project such as specifications, estimates, the schedule, or risks. It is one of the project documents used during the project along with the project management plan and its subsidiary plans. [Initiating and Planning]

PMI®, *PMBOK® Guide*, 2017, 81, 89, 699
PMI® *PMP Examination Content Outline*, 2015, Initiating, 5, Task 2
PMI® *PMP Examination Content Outline*, 2015, Planning, 8, Task 11

13. a. Standard policies, processes, and procedures

Standard policies, processes, and procedures are an organizational process asset to consider as the project management plan is developed. They are one of several organizational process assets that can influence the preparation of the preparation of the project management plan. [Planning]

PMI®, *PMBOK® Guide*, 2017, 84
PMI® *PMP Examination Content Outline*, 2015, Planning, 7, Task 11

14. b. Used to review, approve, or defer change requests

Change control meetings are held on some projects with a change control board. This board reviews the change requests and approves, rejects, or defers them. The meeting assesses the impact of the change. It also may discuss alternatives to consider. Then the decision is communicated to the request owner or group. This board may review configuration items, but it is not mandatory. Configuration management may be handled by a different board. Regardless, roles and responsibilities need to be defined and agreed upon by appropriate stakeholders and are part of the change management plan. [Monitoring and Controlling]

PMI®, *PMBOK® Guide*, 2017, 120
PMI® *PMP Examination Content Outline*, 2015, Monitoring and Controlling, 9, Task 2

15. b. Project management information systems

The items listed are part of these systems, a tool and technique in both processes. In Direct and Manage Project Work, it is a tool and technique providing access to information technology software tools, such as scheduling software tools, work authorization systems, configuration management systems, information collection and distribution systems, and interfaces to other systems in the organization. It may include automated gathering and reporting on key performance indicators. In Manage Program Knowledge, the PMIS also is a tool and technique and often includes a document management system. In Monitor and Control Project Work, it also includes scheduling, cost; and resourcing tools, performance indicators, data bases, project records, and financial information. [Executing and Monitoring and Controlling]

PMI®, *PMBOK® Guide*, 2017, 95, 103, 109
PMI® *PMP Examination Content Outline*, 2015, Executing, 8, Task 2
PMI® *PMP Examination Content Outline*, 2015, Monitoring and Controlling, 9, Task 1

16. b. Updates

Change requests may include corrective actions, preventive actions, defect repairs, or updates. Updates are changes to formally controlled project documents or plans to reflect modified or additional ideas or content. [Executing]

PMI®, *PMBOK® Guide*, 2017, 96
PMI® *PMP Examination Content Outline*, 2015, Executing, 8, Task 4

17. a. Procedures that define how project documents may be changed

A change control system is set of procedures that describes how modifications to the project deliverables and documentation are managed and controlled. Changes may be requested by any stakeholder involved with the project. Change requests may lead to different types of action and may or may not impact the project's baselines. It is necessary to identify changes, document them into a change request, decide on the change, and verify the approved changes are registered, assessed, approved, tracked, and their results are communicated to stakeholders. They may be initiated from inside or outside of the project and may be optional or legally mandated. [Monitoring and Controlling]

PMI®, *PMBOK® Guide*, 2017, 92, 117–119, 700
PMI® *PMP Examination Content Outline*, 2015, Monitoring and Controlling, 9, Task 1

18. d. Project management plan

The project management plan describes how the project will be executed, monitored and controlled, and closed. It integrates and consolidates subsidiary plans and baselines and other information such as project documents needed to manage the project. It also contains other items including information on management reviews to determine if progress is as planned or if preventative or corrective action is needed. [Planning]

PMI®, *PMBOK® Guide*, 2017, 99
PMI® *PMP Examination Content Outline*, 2015, Planning, 7, Task 11

19. d. Describe the business value resulting from the project

The project benefits management plan documents how and when the project benefits will be delivered and the methods to measure the benefits. Benefits are outcomes of actions, behaviors, products, services, or results that provide value to the organization and the project's intended beneficiaries. In developing this plan, it uses the data in the business case, and the cost/benefit analysis in the two documents shows the estimate of the costs compared to the value of the benefits the project will realize. This benefit management plan along with the project management plan describe how the business value of the project is part of the organization's ongoing operations. It includes the metrics to use to verify the business case and the project's success. [Planning]

PMI®, *PMBOK® Guide*, 2017, 33
PMI® *PMP Examination Content Outline*, 2015, Planning, 7, Task 11

20. a. Use visual management tools

    Visual management tools increasingly are used in project integration management rather than written plans and other documents. The purpose is to capture and oversee those aspects of the project considered critical. When key project elements are visible to the entire team and to management, they provide a real-time overview of the project's status, facilitate knowledge transfer, and empower team members and other stakeholder to help identify and resolve issues. [Planning]

    PMI®, *PMBOK® Guide*, 2017, 33
    PMI® *PMP Examination Content Outline,* 2015, Planning, 7, Task 11

21. c. For successful completion of the current and subsequent project phases

    Rolling wave planning provides progressive detailing of the work to be accomplished throughout the life of the project, indicating that planning and documentation are iterative and ongoing processes. Through this approach, where a more general and high-level plan is available, more detailed planning is executed for appropriate time windows as new work activities are to begin, and resources are to be assigned. The work in the near term then is planned in detail, while future work is planned at a higher level. It is especially useful on longer projects. As an example, in decomposing your work breakdown structure (WBS), it may not be possible on large and long projects, and you may need to wait until the project is further along until the complete WBS is finished; rolling wave planning then is used. In defining activities, it is also a tool and technique to plan the near-term activities with the work to be done in the future and planned at a high level. It has similarities to a scaled agile framework. [Planning]

    PMI®, *PMBOK® Guide*, 2017, 160, 185, 721
    PMI® *PMP Examination Content Outline,* 2015, Planning, 7, Task 11

22. d. It provides an integrated perspective

    The purpose of variance analysis is to review the differences for variances between planned and actual performance. It is used throughout project management. In Monitoring and Controlling Project Work, it is used to review the variances from an integrated perspective. To do so cost, time, and resource variances are used in relation to one another. It enables an opportunity to have an overall view of the variance on the project such that the project manager then can determine the appropriate preventive or corrective actions to be taken. [Monitoring and Controlling]

    PMI®, *PMBOK® Guide*, 2017, 111
    PMI® *PMP Examination Content Outline,* 2015, Monitoring and Controlling, 9, Task 2

23. b. Does not apply to the product to be created

    A project is completed when its objectives have been achieved by producing deliverables, or when they are recognized as being unachievable, and the project is terminated, or when the need for the project no longer exists. Thus, the concept of *temporary* applies to the project life cycle—not the product life cycle. Further, temporary does not necessarily mean the project's duration is short as it refers to the project's engagement and its longevity. It recognizes projects have a definite beginning and an end. It also does not typically apply to the product, service, or result as some projects are undertaken to create a lasting outcome with social, economic, material, or environmental impacts that far outlive the project. A national monument is an example of a deliverable that may last for centuries. [Planning]

    PMI®, *PMBOK® Guide*, 2017, 4–5
    PMI® *PMP Examination Content Outline*, 2015, Planning, 7, Task 11

24. d. Review the project management plan

    In closing the project, it is necessary to ensure that the project work is completed, and the project has met its objectives. To ensure this is done, the project manager reviews the project management plan. [Closing]

    PMI®, *PMBOK® Guide*, 2017, 123
    PMI® *PMP Examination Content Outline*, 2015, Closing, 10, Task 3

25. a. Performance measurement baseline

    The performance measurement baseline is an example of an additional component prepared in this process. It integrates the scope, schedule, and cost plan to compare the project work against project execution to measure and integrate performance. Other components are a change management plan, a configuration management plan, the project life cycle, the development approach, and management reviews. The scope baseline is one of three baselines with the other two time and cost. Answers c and d are other documents prepared, which are not part of the project management plan. [Planning]

    PMI®, *PMBOK® Guide*, 2017, 88
    PMI® *PMP Examination Content Outline*, 2015, Planning, 7, Task 11

26. a. Employees who can apply the training effectively

   Most deliverables are tangible, such as buildings or roads, but intangible deliverables also can be provided and may be even more important. Deliverables are any unique or verifiable product, result, or capability to perform a service required to complete a process, phase, of the project. Deliverables can include components of the project management plan. The output of a process often results in an input to another process or a deliverable. Deliverables are outputs of the Direct and Manage Project Work process. It is recommended that change control be applied to the first version of a deliverable, and multiple versions or editions of the deliverable then are controlled and supported by configuration management. [Executing]

   PMI®, *PMBOK® Guide*, 2017, 22, 95 704
   PMI® *PMP Examination Content Outline*, 2015, Executing, 8, Task 2

27. b. Corrective action

   When you volunteered, you signed a confidentiality statement, so you could not disclose what was under way on this activity. Now the Standard has been issued, and to stay in alignment with PMI®'s best practices, you need to issue a change request based on corrective action to realign the performance of the work of your project with your project management plan. [Executing]

   PMI®, *PMBOK® Guide*, 2017, 96
   PMI® *PMP Examination Content Outline*, 2015, Executing, 9, Task 4

28. a. During the initiating processes

   When the project manager is selected and assigned to the project during initiation, several of the usual start-up tasks for a project are simplified. If assigned early, the project manager can participate in development of the project charter. Some project managers are involved even earlier in evaluation and analysis activities. When early involvement is done, the project manager can consult with executives and business leaders on ideas to advance organizational objectives, improve organizational performance, or in meeting customer needs. This early involvement enables the project manager to assist business analysts, help develop the business case, help realize business benefits from the project, and participate in some aspects of portfolio management. [Initiating]

   PMI®, *PMBOK® Guide*, 2017, 51, 77
   PMI® *PMP Examination Content Outline*, 2015, Initiating, 5, Task 6

29. a. Regression analysis

     Regression analysis is a tool and technique in Close Project or Phase under data analysis. It analyzes the interrelationships between different project variables that contributed to the project's outcomes to improve performance on future projects. Other data analysis tools and techniques are document analysis, trend analysis, and variance analysis. [Closing]

     PMI®, *PMBOK® Guide*, 2017, 126
     PMI® *PMP Examination Content Outline*, 2015, Closing, 10, Task 3

30. c. Work performance data

     Work performance data are the raw observations and measurements identified during activities performed to carry out the work of the project. Other examples are the reported percent of work physically completed, quality and technical performance measures, actual costs, and actual durations. It is important to note that the PMBOK® recognizes that the terms data and information are used interchangeably in practice, which can lead to confusion among stakeholders. It has therefore set up a classification system of work performance data, work performance information, and work performance reports. Work performance information is performance data collected in various controlling processes, and work performance reports are physical or electronic representation of work performance information compiled in project documents. [Executing]

     PMI®, *PMBOK® Guide*, 2017, 26, 95
     PMI® *PMP Examination Content Outline*, 2015, Executing, 8, Task 6

31. a. Requirements traceability matrix

     The requirements traceability matrix is an input to the Direct and Manage Project Work process. This matrix links product requirements to the deliverables that satisfy them and helps to focus on the outcome of the project. It helps ensure each requirement has business value since it is linked to the business and organizational objectives. [Executing]

     PMI®, *PMBOK® Guide*, 2017, 93, 148
     PMI® *PMP Examination Content Outline*, 2015, Executing, 8, Task 2

32. a. Confirm the deliverables have been accepted

    Meetings are a tool and technique in Close Project Phase. They are used to confirm the deliverables have been accepted and to validate the exit criteria have been met, and to formalize contract completion. They also are used to evaluate stakeholder satisfaction, to gather lessons learned, to transfer knowledge, and to celebrate success. [Closing]

    PMI®, *PMBOK® Guide*, 2017, 127
    PMI® *PMP Examination Content Outline*, 2015, Closing, 10, Task 1

33. c. A sponsor

    The project charter should be issued by a project initiator or sponsor who formally authorizes the project and provides the project manager with the authority to apply organizational resources to project activities. The sponsor may be a person or a group and is accountable for enabling success, providing resources and support for the project. The project charter should not be issued by the project manager, although, the project manager can assist in its development. [Initiating]

    PMI®, *PMBOK® Guide*, 2017, 81
    PMI® *PMP Examination Content Outline*, 2015, Initiating, 5, Task 6

34. b. Issues are found when work is being performed, but change requests can be submitted to modify documents, deliverables, or baselines

    Issues and change requests are tools and techniques used in Direct and Manage Project Work. Issues can occur during the project's life cycle, as the project manager faces problems, gaps, inconsistencies, and conflicts, and an issue log is used to record and track them. Change requests are used to modify any document, deliverable, or baseline. Change requests are submitted, which can modify policies, procedures, project or product scope, project cost or budget, the project's schedule, or project results. They are processed for review and disposition in the Integrated Change Control Process. They may include corrective or preventive action, defect repair, or updates. [Executing]

    PMI®, *PMBOK® Guide*, 2017, 96
    PMI® *PMP Examination Content Outline*, 2015, Executing, 8, Task 4

35. d. Monitoring and controlling

Monitoring and controlling is an example of a phase that is performed continuously during the project. Many of the processes in it are performed from the start of the project until it is closed. The charter is an example of the output of the Initiating process, which is performed once. Acquiring resources are performed when resources are needed, similar to conducting procurements. [Monitoring and Controlling]

PMI®, *PMBOK® Guide*, 2017, 22
PMI® *PMP Examination Content Outline*, 2015, Monitoring and Controlling, 9, Task 1

36. b. Predictive

A predictive life cycle is recommended when the project's scope, time, and cost to deliver it are determined in the project life cycle as early as possible. Any changes to its scope are monitored carefully. A predictive life cycle may be called a waterfall life cycle. [Planning]

PMI®, *PMBOK® Guide*, 2017, 19
PMI® *PMP Examination Content Outline*, 2015, Planning, 7, Task 11

37. b. Update the closure documents

Even though the project was terminated early, it is the project manager's responsibility to update the project closure documents. The purpose is to indicate why the project was terminated and formalize, if appropriate, any finished or unfinished deliverables to others. [Closing]

PMI®, *PMBOK® Guide*, 2017, 128
PMI® *PMP Examination Content Outline*, 2015, Closing, 10, Task 5

38. a. There is no "paper trail" of change activity

An informal approach means there is a lack of documentation as to the change process and activities. The benefit of the Perform Integrated Change Control process is it allows for documented changes to be considered in an integrated way, and in doing so, it addresses overall project risk, which may arise without considering the overall project objectives or plans. [Monitoring and Controlling]

PMI®, *PMBOK® Guide*, 2017, 113
PMI® *PMP Examination Content Outline*, 2015, Monitoring and Controlling, 9, Task 2

39. a. Measuring customer satisfaction

   The administrative closure procedures involve a number of activities, one of which is to measure stakeholder satisfaction. The other administrative activities involve: activities and actions needed to satisfy completion or exit criteria for the project; activities related to completing contractual agreements applicable to the project; activities necessary to collect project records, audit successes or failures, manage knowledge sharing and transfer, identify lessons learned, and archive information for future use; actions or activities to transfer the result of the project to operations; and collecting suggestions for improving policies and procedures for the organization and sending them to the responsible unit. [Closing]

   PMI®, *PMBOK® Guide*, 2017, 123
   PMI® *PMP Examination Content Outline,* 2015, Closing, 10, Task 7

40. b. Configuration management plan

   A configuration management plan is part of a project management plan. It describes how information and items of the project and which items will be recorded and updated. The purpose is to ensure the product, service, or result of the project are consistent and/or operative. [Planning]

   PMI®, *PMBOK® Guide*, 2017, 88
   PMI® *PMP Examination Content Outline,* 2015, Planning, 7, Task 11

Project Scope Management

Practice Questions

INSTRUCTIONS: Note the most suitable answer for each multiple-choice question in the appropriate space on the answer sheet.

1. Progressive elaboration of product characteristics on your project must be coordinated carefully with the—

   a. Project scope definition
   b. Project stakeholders
   c. Scope change control system
   d. Customer's strategic plan

2. You are examining multiple scope change requests on a project you were asked to take over because the previous project manager decided to resign. To assess the degree to which the project scope will change, you need to compare the requests to which project document?

   a. Preliminary scope statement
   b. WBS
   c. Change management plan
   d. Scope management plan

3. You and your project team recognize the importance of project scope management to a project's overall success; therefore, you include only the work required for successful completion of the project. The first step in the Project Scope Management process is to—

   a. Clearly distinguish between project scope and product scope
   b. Prepare a scope management plan
   c. Define and document your stakeholders' needs to meet the project's objectives
   d. Prepare your requirements documentation

4. An example of an organizational process asset that could affect how project scope is to be managed is—

   a. Personnel administration
   b. Marketplace conditions
   c. Historical information
   d. Organizational culture

5. You are managing a complex project for a new method of heating and air conditioning in vehicles. You will use both solar and wind technologies in this project to reduce energy costs. Therefore, you must ensure that the work of your project will result in delivering the project's specified scope, which means that you should measure completion of the product scope against the—

   a. Scope management plan
   b. Business requirements
   c. Solution requirements
   d. Requirements management plan

6. A key tool and technique used in Define Scope is—

   a. Templates, forms, and standards
   b. Decomposition
   c. Multi-criteria decision analysis
   d. Project management methodology

7. Each organization is unique and has different ways to define its products, services, and results. You are working on your project to develop a new product using disruptive technologies to reduce traffic in your city. You are working to define your project's scope, and you have a large team plus interested stakeholders. A tool and technique that may be helpful is—

   a. Sensitivity analysis
   b. Decision trees
   c. Facilitation
   d. Lateral thinking

8. Assume you and your team prepared your project's scope statement, and it was approved by your sponsor and your governance committee. However, you also need to update some project documents as an output of the Define Scope process. One example is—

   a. Value engineering
   b. Assumptions log
   c. WBS Dictionary
   d. Change log

9. The document to evaluate whether requests for changes or additional work are contained within or outside the project's exclusion is provided by the—

   a. Project management plan
   b. Project scope statement
   c. Project scope management plan
   d. WBS dictionary

10. Rather than use a WBS, your team developed a bill of materials to define the project's work components. A customer review of this document uncovered that a scope change was needed, because a deliverable had not been defined, and it led to the need to write a change request. This is an example of a change request that is required because it is the result of—

    a. An external event
    b. A change to the scope baseline
    c. A value-adding change
    d. An error or omission in defining the scope of the project

11. Collecting requirements is critical in project scope management, but all the requirements may not be used as the project scope is defined. Assume once you defined your project's scope, you then began to develop your work breakdown structure. As you worked on it, you found it useful to review the—

    a. Scope management plan
    b. Requirements documentation
    c. Requirements management plan
    d. Requirements traceability matrix

12. Assume you are preparing the scope statement for your project. In it, you are describing the project's deliverables, assumptions, and constraints. You also want to include—

    a. Project exclusions
    b. The business need
    c. Stakeholder expectations
    d. Project management methodology

13. Assume you are working to collect the requirements for your project. As you do so, you realize your project's success is due to—

    a. Active stakeholder involvement
    b. Linking the requirements you collect to the requirement management plan
    c. Conducting facilitated workshops with stakeholders
    d. Preparing a requirements document template that you and your team can use throughout the Collect Requirements process

14. You want to structure your project so that each project team member has a discrete work package to perform. The work package is a—

    a. Work defined at the lowest level of the WBS
    b. Task with a unique identifier
    c. Required level of reporting
    d. Task that can be assigned to more than one organizational unit

15. Quality function deployment is one approach for collecting requirements. Assume that you have studied the work of numerous quality experts, such as Deming, Juran, and Crosby, and your organization has a policy that states the importance of quality as the key constraint of all project constraints. You and your team have decided to use quality function deployment on your new project to manufacture turbines that use alternative fuels. The first step you should use is to—

    a. Determine the voice of the customer
    b. Build the house of quality
    c. Address the functional requirements and how best to meet them
    d. Hold a focus group of prequalified stakeholders

16. On the WBS, you have decided to structure it by phases of your project. In this approach—

    a. Deliverables are at the third level
    b. Subcomponents are at the first level
    c. Major deliverables are at the second level
    d. Project organizational units are at the third level

17. Change is inevitable on projects. The number of interested stakeholders, technology, the difficulty in obtaining resources, and the difficulty in motivating people to be on the team are just a few changes that may occur on projects. Projects also cause changes in the development of a new product, service, or result. The project manager wants to minimize changes on the project. Uncontrolled changes are often referred to as—

    a. Rework
    b. Scope creep
    c. Configuration items
    d. Emergency changes

18. You are going to be using a number of contractors on your project to produce a car that uses natural gas as its fuel supply since your country has an abundance of it available. You are working on preparing your WBS for this project as you have completed the scope statement. In your WBS, the most effective approach to use is to—

    a. Link the WBS to the bill of materials
    b. Integrate subcomponents
    c. Use your organization's template
    d. Focus on major deliverables

19. In Control Scope it is important to determine the cause of any unacceptable variance relative to the scope baseline. This can be done through—

    a. Root-cause analysis
    b. Control charts
    c. Inspections
    d. Data analysis

20. To assist your software development team in collecting requirements from potential users and to ensure that agreement about the stakeholders' needs exists early in the project, you decide to use an interpersonal and team skill. Numerous techniques are available, but you and your team choose a voting process to rank the most useful ideas for further prioritization. This approach is known as—

    a. Brainstorming
    b. Nominal group technique
    c. Delphi technique
    d. Affinity diagram

21. You have been appointed project manager for a new project in your organization and must prepare a project management plan. You decide to prepare a WBS to show the magnitude and complexity of the work involved. No WBS templates are available to help you. To prepare the WBS, your first step should be to—

    a. Determine the cost and duration estimates for each project deliverable
    b. Identify and analyze the deliverables and related work
    c. Identify the components of each project deliverable
    d. Determine the key tasks to be performed

22. You want to avoid scope creep on your project and are working hard to do so. Your sponsor has asked for regular reports as to how the project is performing according to the scope baseline. You decide to use—

    a. Variance analysis
    b. Inspection and peer review results
    c. Work performance information
    d. The impact analysis results of proposed change requests

23. You are leading a project team to identify potential new products for your organization. One idea was rejected by management because it would not fit with the organization's core competencies. You now have a project to make your organization one that is comfortable with using disruptive technologies, which was approved. Now you are working on your scope statement and as you do so, it is helpful to consider—

    a. The assumption log
    b. Project life cycle definition
    c. Marketplace conditions
    d. Development approach

24. Validate scope—

    a. Improves cost and schedule accuracy, particularly on projects using innovative techniques or technology
    b. Is the last activity performed on a project before handoff to the customer
    c. Documents the characteristics of the product, service, or result that the project was undertaken to create
    d. Differs from Perform Quality Control in that Validate Scope is concerned with the acceptance—not the correctness—of the deliverables

25. Although you and your team are actively working to control the scope of your project, change is inevitable on projects. You have had some change requests from your sponsor, team members, and other stakeholders. You have processed these change requests. This means you also need to update which of the following documents?

    a. Scope management plan
    b. Requirements traceability matrix
    c. WBS
    d. Scope statement

26. One approach that can be used to detect the impact of any change from the scope baseline on the project objectives is—

    a. Requirements traceability matrix
    b. Review of formal and informal scope control processes and guidelines
    c. A formal configuration management plan
    d. Well-documented requirements

27. Throughout the project, you are focusing on the Monitoring and Controlling process, which include Control Scope. You have processed some change requests and now also need to update some parts of the project management plan, one of which is the—

    a. Requirements management plan
    b. Lessons-learned register
    c. Cost baseline
    d. Requirements documents

28. Validate Scope is performed throughout the project. However, there are several outputs from the Validate Scope process, one of which is—

    a. Inspection reports
    b. Voting results
    c. Work performance data
    d. Deliverables that have not been formally accepted

29. Your project is now under way, and you are working with your team to prepare your requirements management plan. One component of this plan you want to include is—

    a. Metrics to use
    b. A set of procedures by which project scope and product scope may be changed
    c. Requirements traceability matrix
    d. Requirements documentation

30. You are the project manager on a systems engineering project designed to last six years and to develop the next-generation corvette for use in military operations. You and your team recognize that requirements may change as new technologies, especially in sonar systems, are developed. You are concerned that these new technologies may lead to changes in the scope of your product, which then will affect the scope of your project. Therefore to help you, you should prepare a—

    a. Requirements traceability matrix
    b. Requirements prioritization process
    c. Configuration management system
    d. Requirements management plan

31. Your customer signed off on the requirements document and scope statement of your video game project last month. Today she stated she would like to make it an interactive game that can be played on a television, smart phone. and on a computer. This represents a requested scope change that at a minimum—

    a. Should be reviewed according to the Perform Integrated Change Control process
    b. Results in a change to all project baselines
    c. Requires adjustments to cost, time, quality, and other objectives
    d. Results in a lesson learned

32. While the Validate Scope process has some similarities with the Control Quality process, a key benefit of Validate Scope is—

    a. It formalizes acceptance of the project's deliverables
    b. It decreases the number of change requests
    c. It brings objectivity to the acceptance process
    d. It reduces the need for product reviews

33. Modifications may be needed to the WBS and WBS dictionary because of approved change requests, which shows that—

    a. Replanning is an output of Control Scope
    b. Scope creep is common on projects
    c. Rebaselining will be necessary
    d. Updates are needed to the scope baseline

34. You and your team are documenting requirements on your project to control fatigue as people need to work more hours to keep up with the competition. You decided to set up categories for the requirements on your project. Temporary capabilities are an example of—

    a. Stakeholder requirements
    b. Transition requirements
    c. Project requirements
    d. Business requirements

35. You are working to collect the requirements for your project to eliminate the possibility of later scope creep. You have a variety of tools and techniques you can use. Assume you want to obtain early feedback on the requirements, and you have decided the most appropriate tool and technique is—

    a. Interviews
    b. Prototypes
    c. Brainstorming
    d. Requirements management plan

36. You have prepared the WBS for your project. However, for your project, your company decided to try an agile approach. Many feel you will not need to develop a WBS and a lot of other documents in your project management methodology. However, with agile the difference is—

    a. Different levels of decomposition can be used
    b. Epics are decomposed into user stories
    c. The WBS is represented as a hierarchical breakdown
    d. The project management element is not included because of the close customer collaboration

37. The project scope statement is important in scope control because it—

    a. Is a critical component of the scope baseline
    b. Provides information on project performance
    c. Alerts the project team to issues that may cause problems in the future
    d. Is expected to change throughout the project

38. The product scope description is documented as part of the project's scope statement. It is important to include it because it—

    a. Facilitates the project acceptance process
    b. Describes specific constraints associated with the project
    c. Progressively elaborates characteristics
    d. Shows various alternatives considered

39. How is a context diagram used?

    a. To depict product scope
    b. To trace requirements as part of the traceability matrix
    c. To develop the scope management plan
    d. To develop the requirements management plan

40. You are establishing a PMO that will have a project management information system that will be an online repository of all program data. You will collect descriptions of all work components for each project under the PMO's jurisdiction. This information will form an integral part of the—

    a. Chart of accounts
    b. WBS dictionary
    c. WBS structure template
    d. Earned value management reports

Answer Sheet

1.	a	b	c	d		21.	a	b	c	d
2.	a	b	c	d		22.	a	b	c	d
3.	a	b	c	d		23.	a	b	c	d
4.	a	b	c	d		24.	a	b	c	d
5.	a	b	c	d		25.	a	b	c	d
6.	a	b	c	d		26.	a	b	c	d
7.	a	b	c	d		27.	a	b	c	d
8.	a	b	c	d		28.	a	b	c	d
9.	a	b	c	d		29.	a	b	c	d
10.	a	b	c	d		30.	a	b	c	d
11.	a	b	c	d		31.	a	b	c	d
12.	a	b	c	d		32.	a	b	c	d
13.	a	b	c	d		33.	a	b	c	d
14.	a	b	c	d		34.	a	b	c	d
15.	a	b	c	d		35.	a	b	c	d
16.	a	b	c	d		36.	a	b	c	d
17.	a	b	c	d		37.	a	b	c	d
18.	a	b	c	d		38.	a	b	c	d
19.	a	b	c	d		39.	a	b	c	d
20.	a	b	c	d		40.	a	b	c	d

Answer Key

1. a. Project scope definition

   Progressive elaboration of a project's specification must be coordinated carefully with scope definition. It elaborates the characteristics of the product, service, or result as defined in the charter and requirements documents. The scope statement, the major output of the Define Scope process, describes the project's scope, major deliverables, and constraints. This means it includes both project and product scope. [Planning]

   PMI®, *PMBOK® Guide*, 2017, 154
   PMI® *PMP Examination Content Outline*, 2015, Planning, 6, Task 2

2. b. WBS

   The WBS, along with the approved scope statement and the WBS dictionary, defines the project's scope baseline, which provides the basis for any changes that may occur on the project. Formal change control procedures should be used if the scope baseline requires change. It is a component of the project management plan. [Planning]

   PMI®, *PMBOK® Guide*, 2017, 161
   PMI® *PMP Examination Content Outline*, 2015, Planning, 6, Task 2

3. b. Prepare a scope management plan

   The work involved in the six Project Scope Management processes begins by preparing a scope management plan, which is a subsidiary plan for the project management plan. It documents how project scope will be defined, developed, monitored, controlled, and validated and describes the Project Scope Management processes from collecting requirements to control and provides guidance as to how scope will be managed throughout the project. [Planning]

   PMI®, *PMBOK® Guide*, 2017, 131
   PMI® *PMP Examination Content Outline*, 2015, Planning, 6, Task 2

4. c. Historical information

   Organizational process assets that can influence Plan Scope Management include formal and informal policies, procedures, and guidelines impacting project scope management. The lessons-learned repository is another example of an organizational process asset. The other answers are examples of enterprise environmental factors to consider. [Planning]

   PMI®, *PMBOK® Guide*, 2017, 136
   PMI® *PMP Examination Content Outline*, 2015, Planning, 6, Task 2

5. c. Requirements management plan

Completion of the project scope is measured against the project management plan, and completion of the product scope is measured against the requirements. The requirements documents are an output from the Collect Requirements process, and requirements may be further classified in terms of business requirements, stakeholder requirements, solution requirements, transition and readiness requirements, project requirements, and quality requirements. Solution requirements are further classified into technical requirements, which are functional requirements related to the product, and nonfunctional requirements, which describe the environmental conditions needed for the product to be effective. [Planning]

PMI®, *PMBOK® Guide*, 2017, 148
PMI® *PMP Examination Content Outline*, 2015, Planning, 6, Task 1

6. c. Multi-criteria decision analysis

The key purpose of the Define Scope process is to prepare the project scope statement. In doing so, a number of tools and techniques can be used, one of which is multi-criteria analysis, part of the decision analysis tool and technique. It is a technique that uses a decision matrix. The decision matrix provides a systematic analytical approach to establish criteria, and it may include requirements, schedule, budget, and resources. The purpose is to help refine the product and project scope. [Planning]

PMI®, *PMBOK® Guide*, 2017, 153
PMI® *PMP Examination Content Outline*, 2015, Planning, 6, Task 2

7. c. Facilitation

It is easy to overlook the need for interpersonal and team skills especially in processes such as Define Scope. The key words in this question are a large team and interested stakeholders. In the Define Scope process, an interpersonal and team skill that is used is facilitation. It is used in workshops and working sessions with key stakeholders, and each stakeholder may have different expectations or areas of expertise, which can be beneficial. Facilitation then can help obtain a cross functional and common understanding among the stakeholders concerning project deliverables and any project or product boundaries. [Planning]

PMI®, *PMBOK® Guide*, 2017, 153
PMI® *PMP Examination Content Outline*, 2015, Planning, 6, Task 2

8. b. Assumptions log

The assumptions log is both an input to the Define Scope process and an output from it as a project document to update. As an input, it is a project document developed in Integration Management that identifies both assumptions and constraints about the project product, environment, stakeholders, and other factors that may influence product and project scope. As the scope statement is prepared, these assumptions and constraints are reviewed, and after the scope statement is finalized and approved, this log is an example of other project documents to update as ideas about assumptions and constraints may have changed during this process. Additional assumptions and constraints may be identified, and others originally identified may no longer be ones that will affect the product or project. [Planning]

PMI®, *PMBOK® Guide,* 2017, 165
PMI® *PMP Examination Content Outline,* 2015, Planning, 6, Task 2

9. b. Project scope statement

Sections of the project's scope statement include the product scope description, deliverables, acceptance criteria, and project exclusions. Project exclusions identify generally what is excluded from the project, and states explicitly what is excluded from the project to help manage stakeholder expectations and to reduce scope creep. [Planning]

PMI®, *PMBOK® Guide,* 2017, 154
PMI® *PMP Examination Content Outline,* 2015, Planning, 6, Task 2

10. b. A change to the scope baseline

The scope baseline consists of the scope statement, the WBS, and the WBS Dictionary. Since a Bill of Materials was used instead of a WBS, a change request is required since a Bill of Materials provides a hierarchical view of the physical assemblies, subassemblies, and components needed to build a manufactured product. The WBS instead is a deliverable-oriented grouping of project components used to define the total scope of the project, providing a structured vision of what has to be delivered. The purpose of the Control Scope process is to monitor the project's status and that of the product and manage changes to the scope baseline. Its benefit is the scope baseline then is maintained during the project. Using a bill of materials where a WBS would be more appropriate may result in an ill-defined scope and subsequent change requests. [Monitoring and Controlling]

PMI®, *PMBOK® Guide,* 2017, 167
PMI® *PMP Examination Content Outline,* 2015, Monitoring and Controlling Planning, 9, Task 2

11. b. Requirements documentation

In preparing the WBS, it is helpful to review two project documents, one is the requirements documentation as the detailed requirements define how individual requirements meet the project's business need. The other document to review is the project scope statement. [Planning]

PMI®, *PMBOK® Guide*, 2017, 157
PMI® *PMP Examination Content Outline*, 2015, Planning, 6, Task 2

12. a. Project exclusions

The purpose of the scope statement is to describe the project scope, deliverables, assumptions, and constraints, and it documents the entire project and product scope. It also provides a common understanding among stakeholders. In addition, it contains project exclusions or what is excluded from the project. They are included to state explicitly what is out of scope for the project, which then helps to manage stakeholder expectations and to reduce scope creep. [Planning]

PMI®, *PMBOK® Guide*, 2017, 154
PMI® *PMP Examination Content Outline*, 2015, Planning, 6, Task 2

13. a. Active stakeholder involvement

Collecting requirements involves determining, documenting, and managing stakeholder needs and requirements to meet the project's objectives. As you work to collect requirements, active stakeholder involvement is critical to overall success in the discovery and decomposition of needs into both project and product requirements. It also involves taking care in determining, documenting, and managing requirements of the project's product, service, or result. [Planning]

PMI®, *PMBOK® Guide*, 2017, 113
PMI® *PMP Examination Content Outline*, 2015, Planning, 7, Task 13

14. a. Work defined at the lowest level of the WBS

A work package is the work defined at the lowest level of the WBS. It is used to estimate and manage the cost and duration of the work defined at this level. [Planning]

PMI®, *PMBOK® Guide*, 2017, 158
PMI® *PMP Examination Content Outline*, 2015, Planning, 6, Task 2

15. a. Determine the voice of the customer

   Facilitation is a tool and technique in Collect Requirements. These are focused sessions with stakeholders to define product requirements. If conducted appropriately, they can quickly define functional requirements and reconcile stakeholder differences in an interactive way. Quality function development is an example of a facilitation technique as it is used to help determine critical characteristics for new product development. It begins by collecting customer needs and is also known as the voice of the customer. The next step is to sort these needs and prioritize them, so goals can be set to achieve them. [Planning]

   PMI®, *PMBOK® Guide*, 2017, 145
   PMI® *PMP Examination Content Outline*, 2015, Planning, 6, Task 2

16. a. Deliverables are at the third level

   The WBS includes all work needed to be done to complete the project. The WBS structure can be set up using different approaches. If a phase approach is used, the phases of the project life cycle are shown at the second level of decomposition with product and project deliverables at the third level. [Planning]

   PMI®, *PMBOK® Guide*, 2017, 159
   PMI® *PMP Examination Content Outline*, 2015, Planning, 6, Task 2

17. b. Scope creep

   Project scope creep is typically the result of uncontrolled changes in product or project scope without adjustments to time, cost, or resources. Scope control works to control the impact of any project scope changes, manage actual changes when they occur, and it is integrated with other control processes. Scope control then ensures all requested changes and recommended corrective or preventive actions are processed through the Perform Integrated Change Control process. [Monitoring and Controlling]

   PMI®, *PMBOK® Guide*, 2017, 168
   PMI® *PMP Examination Content Outline*, 2015, Monitoring and Controlling, 9, Task 2

18. b. Integrate subcomponents

The key words in this question are a number of contractors. This means the WBS is developed by integrating subcomponents that are developed from outside of the project team, such as the use of contractors. The contractor or seller then develops its WBS from the supporting contract WBS as part of the contracted work. The various subcomponents are then incorporated to form the project's WBS. [Planning]

PMI®, *PMBOK® Guide*, 2017, 159
PMI® *PMP Examination Content Outline*, 2015, Planning, 6, Task 2

19. d. Data analysis

Data analysis is the tool and technique for Control Scope. It consists of variance analysis and trend analysis. Variance analysis compares the baseline to the actual results and whether the variance is within the accepted threshold. Its purpose is to see whether corrective or preventive action is needed. Trend analysis evaluates project performance over time to assess if performance is improving or is deteriorating. [Monitoring and Controlling]

PMI®, *PMBOK® Guide*, 2017, 170
PMI® *PMP Examination Content Outline*, 2015, Monitoring and Controlling, 9, Task 1

20. b. Nominal group technique

The nominal group technique enhances brainstorming with a voting process, which is used to rank the most useful ideas for further brainstorming or for prioritization. It is an interpersonal and team skill as a tool and technique in Collect Requirements. In this approach, people who are participating receive a question or problem, and each person in silence generates and writes down his or her ideas. Next, the moderator writes down the ideas and records them typically on a flip chart. The people involved then discuss each recoded idea until they understand it. Finally, people vote privately to prioritize the ideas and may use a scale such as one to five, where one has the lowest interest, and a five has the highest interest. Voting follows and may take several rounds to reduce the number of ideas and focus on the ones with the highest votes. After each round, the moderator tallies the votes, and the ideas with the highest scores are selected. [Planning]

PMI®, *PMBOK® Guide*, 2017, 144–145
PMI® *PMP Examination Content Outline*, 2015, Planning, 7, Task 13

21. b. Identify and analyze the deliverables and related work

Identifying and analyzing the deliverables and related work is the first step in the decomposition of a project, a tool and technique in the Create WBS process. Decomposition divides and subdivides the scope and deliverables into smaller parts that are more manageable. The deliverables should be defined in terms of how the project will be organized. For example, the major project deliverables may be used as the second level. Once the deliverables and related work are identified, the other steps in decomposition are to structure and organize the related work, decompose upper WBS levels into lower-level detailed components, develop and assign identification codes to the WBS components, and verify the degree of decomposition of the deliverable is appropriate. [Planning]

PMI®, *PMBOK® Guide*, 2017, 160
PMI® *PMP Examination Content Outline*, 2015, Planning, 6, Task 2

22. c. Work performance information

Work performance information is an output of the Control Scope process. It includes correlated and contextualized information on how the project scope is performing against the scope baseline. It may include categories of changes received, identified scope variances and their causes, the impact on schedule or cost, and a forecast of future scope performance. [Monitoring and Controlling]

PMI®, *PMBOK® Guide*, 2017, 170
PMI® *PMP Examination Content Outline*, 2015, Monitoring and Controlling, 9, Task 2

23. a. The assumption log

Assumptions are factors that for planning purposes are considered to be true, real, or certain without proof or demonstration. They are listed in a separate log, which is one of the project documents that is an input to Define Scope. This log contains both assumptions and constraints concerning the product, project environment, and other factors that may influence the scope of the product or the project. This log is used throughout the project and is updated often. [Planning]

PMI®, *PMBOK® Guide*, 2017, 152
PMI® *PMP Examination Content Outline*, 2015, Planning, 6, Task 2

24. d. Differs from Perform Quality Control in that Validate Scope is concerned with the acceptance—not the correctness—of the deliverables

    Documentation that the customer or the sponsor has accepted completed deliverables is an output of Validate Scope. Acceptance criteria are formally signed off and approved by the customer or sponsor. The formal documentation also is received from the customer or sponsor acknowledging formal stakeholder acceptance of the deliverables of the project. [Monitoring and Controlling]

    PMI®, *PMBOK® Guide*, 2017, 166
    PMI® *PMP Examination Content Outline*, 2015, Monitoring and Controlling, 9, Task 1

25. b. Requirements traceability matrix

    Updates to project documents is an output of the Control Scope process. One of these documents is the requirements traceability matrix. It may need updates to reflect changes in requirements documents. [Monitoring and Controlling]

    PMI®, *PMBOK® Guide*, 2017, 171
    PMI® *PMP Examination Content Outline*, 2015, Monitoring and Controlling, 9, Task 2

26. a. Requirements traceability matrix

    It is an input to the Control Scope process as it helps to detect the impact of any change or deviation from the scope baseline to the project objectives. It also may provide the status of any requirements being considered. [Monitoring and Controlling]

    PMI®, *PMBOK® Guide*, 2017, 169
    PMI® *PMP Examination Content Outline*, 2015, Monitoring and Controlling, 9, Task 2

27. c. Cost baseline

The cost baseline is included in the project management plan. It may need updates as part of the Control Scope process. Changes in the cost baseline are incorporated if there are approved scope changes, as in this question. Changes also may be needed to the cost baseline if there are changes in resources or cost estimates, both of which may be necessary from the scope changes. It some cases while controlling the scope, the cost variances that result may be so severe that a revised cost baseline is needed for a realistic basis for performance measurement. [Monitoring and Controlling]

PMI®, *PMBOK® Guide*, 2017, 171
PMI® *PMP Examination Content Outline*, 2015, Monitoring and Controlling, 9, Task 2

28. d. Deliverables that have not been formally accepted

In the Validate Scope process, change requests are an output. They include documentation of the completed deliverables that have not been documented and the reasons why they were not accepted. These deliverables then may be ones that require a change request for defect repair. The change requests are processed through the Integrated Change Control process. [Monitoring and Controlling]

PMI®, *PMBOK® Guide*, 2017, 166
PMI® *PMP Examination Content Outline*, 2015, Monitoring and Controlling, 9, Task 1

29. a. Metrics to use

The requirements management plan defines how requirements will be analyzed, documented, and managed. It includes among other items the metrics to be used and the rationale for using each of them. Some organizations refer to this plan as a business analysis plan. [Planning]

PMI®, *PMBOK® Guide*, 2017, 137
PMI® *PMP Examination Content Outline*, 2015, Planning, 7, Task 11

30. a. Requirements traceability matrix

   The requirements traceability matrix is an output of the Collect Requirements process. It is a grid that links requirements from their origin to the deliverables that satisfy them. It helps ensure the requirements add business value and are not just nice to have features as each requirement is linked to the business and project objectives. It also tracks the requirements throughout the project life cycle. [Planning]

   PMI®, *PMBOK® Guide*, 2017, 148
   PMI® *PMP Examination Content Outline*, 2015, Planning, 7, Task 11

31. a. Should be reviewed according to the Perform Integrated Change Control process

   A requested change is an output from the Control Scope process. Such a change should be handled according to the Perform Integrated Change Control process and may result in an update to the scope baseline and schedule baselines or other components of the project management plan. [Monitoring and Controlling]

   PMI®, *PMBOK® Guide*, 2017, 170
   PMI® *PMP Examination Content Outline*, 2015, Monitoring and Controlling, 9, Task 1

32. c. It brings objectivity to the acceptance process

   There are two key benefits of the Validate Scope process. One is that it brings objectivity to the acceptance process. It also increases the probability of final product, service, or result acceptance as it validates each deliverable. It is performed throughout the project as needed. Its purpose is one of the other possible answers in that it formalizes acceptance of the completed project deliverables. [Monitoring and Controlling]

   PMI®, *PMBOK® Guide*, 2017, 163
   PMI® *PMP Examination Content Outline*, 2015, Monitoring and Controlling, 9, Task 1

33. d. Updates are needed to the scope baseline

If the approved change requests have an effect on the project scope, the WBS, and the WBS dictionary, and the scope statement need to be revised ad reissued to reflect the approved changes following the Perform Integrated Change Control process. The schedule baseline also may require revision plus some other components of the project management plan. [Monitoring and Controlling]

PMI®, *PMBOK® Guide*, 2017, 170
PMI® *PMP Examination Content Outline*, 2015, Monitoring and Controlling, 9, Task 2

34. b. Transition requirements

There are a number of different ways to categorize requirements. This question is an example of transition and readiness requirements. They are temporary capabilities needed to transition from the current or as is stated to the desired future state. Examples include data conversion and training requirements. [Planning]

PMI®, *PMBOK® Guide*, 2017, 148
PMI® *PMP Examination Content Outline*, 2015, Planning, 7, Task 11

35. b. Prototypes

Prototypes are useful to obtain early feedback on requirements. They provide a model of the expected product before it is built. Then, stakeholders can experiment with the model for the final product instead of discussing abstract requirements for it. This approach supports progressive elaboration with iterative cycles of mock-ups, user experiments, feedback, and prototype revision. [Planning]

PMI®, *PMBOK® Guide*, 2017, 147
PMI® *PMP Examination Content Outline*, 2015, Planning, 7, Task 11

36. b. Epics are decomposed into user stories

A WBS is prepared using agile with the difference being that epics are decomposed into user stories. An epic is a large amount of work with a common objective with examples a customer request or a feature. It serves to define user needs. Specific details then are the purpose of user stories. [Planning]

PMI®, *PMBOK® Guide*, 2017, 160
PMI® *PMP Examination Content Outline*, 2015, Planning, 7, Task 11

37. a. Is a critical component of the scope baseline

The project scope statement, along with the WBS, planning package, and WBS dictionary, is a key input to Control Scope. They form the scope baseline, which is compared to actual results to determine if a change, corrective action, or preventive action is needed. [Monitoring and Controlling]

PMI®, *PMBOK® Guide*, 2017, 169
PMI® *PMP Examination Content Outline*, 2015, Monitoring and Controlling, 9, Task 1

38. c. Progressively elaborates characteristics

The project scope statement describes the deliverables and the work required to create them. It contains both product and project scope. It also provides a common understanding of the scope among stakeholders. The product scope statement is a key component as it expands on the level of detail in the project charter. They are both progressively elaborated during the project. [Planning]

PMI®, *PMBOK® Guide*, 2017, 154
PMI® *PMP Examination Content Outline*, 2015, Planning, 7, Task 11

39. a. To depict product scope

It is a tool and technique in Collect Requirements and is an example of a scope model. The context diagram visually depicts the product scope as it shows a business system (process, equipment, or computer, etc.) and how people and other systems (actors) interact with it. The diagram shows inputs to the business system, the people providing the inputs, outputs from the business system, and people receiving the output. [Planning]

PMI®, *PMBOK® Guide*, 2017, 146
PMI® *PMP Examination Content Outline*, 2015, Planning, 7, Task 11

40. b. WBS dictionary

The WBS dictionary typically includes a code of accounts identifier, a description of work, assumptions and constraints, responsible organization, a list of schedule milestones, associated schedule activities, required resources, cost estimates, quality requirements, acceptance criteria, technical references, and any additional information. It provides detailed deliverable, activity, and scheduling information about each component in the overall WBS dictionary of each project's scope baseline. [Planning]

PMI®, *PMBOK® Guide*, 2017, 162
PMI® *PMP Examination Content Outline*, 2015, Planning, 7, Task 11

Project Schedule Management

Practice Questions

INSTRUCTIONS: Note the most suitable answer for each multiple-choice question in the appropriate space on the answer sheet.

Use the following network diagram to answer questions 1 through 4. Activity names and duration are provided.

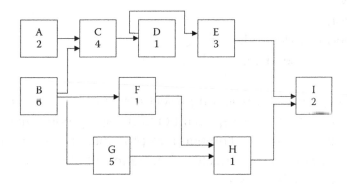

1. What is the duration of the critical path in this network?

   a. 10
   b. 12
   c. 14
   d. 15

2. What is the float for Activity G?

   a. −2
   b. 0
   c. 1
   d. 4

3. If a project planner imposes a finish time of 14 on the project with no change in the start date or activity durations, what is the total float of Activity E?

   a. –1
   b. 0
   c. 2
   d. Cannot be determined

4. If the imposed finish time in question 3 above is removed and reset to 16 and the duration of Activity H is changed to 3, what is the late finish for Activity G?

   a. –11
   b. 11
   c. –13
   d. 13

5. Your company, which operates one of the region's largest chemical processing plants, has been convicted of illegally dumping toxic substances into the local river. The court has mandated that the required cleanup activities be completed by February 15. This date is an example of—

   a. A key event
   b. A milestone
   c. A discretionary dependency
   d. An external dependency

6. You are managing a construction project for a new city water system. The contract requires you to use special titanium piping equipment that is guaranteed not to corrode. The titanium pipe must be resting in the ground a total of 10 days before connectors can be installed. In this example, the 10-day period is defined as—

   a. Lag
   b. Lead
   c. Float
   d. Slack

7. You are working to control your schedule. You have a number of tools and techniques you can use and may need to use several of them for effectiveness. You decide you want to track work remaining to be done so you will use—

   a. Iteration burndown charts
   b. Milestone charts
   c. Applied schedule reserves
   d. Calculation of multiple work package durations

8. You are planning to conduct the team-building portion of your new project management training curriculum out-of-doors in the local park. You are limited to scheduling the course at certain times of the year, and the best time for the course to begin is mid-July. You are using the precedence diagramming method for your schedule. In it, you can show relationships between tasks in four ways, and the most common is—

   a. Finish-to-start
   b. Finish-to-finish
   c. Start-to-start
   d. Start-to-finish

9. Project schedule development is an iterative process. If the start and finish dates are not realistic, the project probably will not finish as planned. You are working with your team to define how to monitor any schedule changes. You documented your decisions in which of the following?

   a. Schedule change control procedures
   b. Schedule management plan
   c. Schedule risk plan
   d. Service-level agreement

10. If, when developing your project schedule, you want to define a distribution of probable results for each schedule activity and use that distribution to calculate another distribution of probable results for the total project, the most common technique to use is—

    a. What-if scenario analysis
    b. Monte Carlo analysis
    c. Linear programming
    d. Concurrent engineering

11. Your lead engineer estimates that a work package will most likely require 50 weeks to complete. It could be completed in 40 weeks if all goes well, but it could take 180 weeks in the worst case. What is the expected duration of the work package?

    a. 45 weeks
    b. 70 weeks
    c. 90 weeks
    d. 140 weeks

12. Your customer wants the project to be completed six months earlier than planned. You believe you can meet this target by overlapping project activities. The approach you plan to use is known as—

    a. Critical chain
    b. Fast tracking
    c. Resource leveling
    d. Crashing

13. Activity A has a duration of three days and begins on the morning of Monday the 4th. The successor activity, B, has a finish-to-start relationship with A. The finish-to-start relationship has three days of lag, and activity B has a duration of four days. Sunday is a non-workday. Such data can help to determine—

    a. The total duration of both activities is 8 days
    b. Calendar time between the start of A to the finish of B is 11 days
    c. The finish date of B is Wednesday the 13th
    d. Calendar time between the start of A to the finish of B is 14 days

14. You can use various estimating approaches to determine activity durations. When you have a limited amount of information available about your project, especially when in the early phases, the best approach to use is—

    a. Bottom-up estimating
    b. Analogous estimating
    c. Reserve analysis
    d. Parametric analysis

15. "I cannot test the software until I code the software." This expression describes which of the following dependencies?

    a. Internal
    b. Rational
    c. Preferential
    d. Mandatory or hard

16. Working with your team to provide the basis for measuring and reporting schedule progress, you agree to consider the—

    a. Schedule data
    b. Network diagram
    c. Schedule baseline
    d. Technical baseline

17. Your project schedule was approved. Management has now mandated that the project be completed as soon as possible. However, you do not think it is possible given resource constraints. In order to convince your management of your need for additional resources, you decide to use—

    a. Resource manipulation
    b. Resource breakdown structure
    c. Critical chain scheduling
    d. Resource leveling

18. Review the following network diagram and table. Of the various activities, which ones would you crash and in what order?

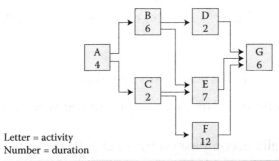

Letter = activity
Number = duration

	Time required, weeks		Cost $		Crashing cost per weeks, $
Activity	Normal	Crash	Normal	Crash	
A	4	2	10,000	14,000	2000
B	6	5	30,000	42,500	12,500
C	2	1	8000	9500	1500
D	2	1	12,000	18,000	6000
E	7	5	40,000	52,000	6000
F	12	3	20,000	29,000	3000
G	6	2	5000	30,000	6000

    a. A, C, E, and F
    b. A, B, D, and F
    c. A, B, E, and F
    d. C, A, F, and G

19. You are working on a project to remodel a home, and you and your team decided you should first finish the plumbing work and then do the electrical work. In this example, this is an example of a—

    a. Start-to-finish approach
    b. Discretionary dependency
    c. Mandatory dependency
    d. Finish-to-start approach

20. Decomposition is a technique used for both WBS development and activity definition. However, in Define Activities, decomposition—

    a. The final output is described in terms of work packages in the WBS.
    b. The final output is described as deliverables or tangible items.
    c. The final output is described as activities.
    d. Is used the same way in scope definition and activity definition.

21. When sequencing project activities in the schedule, you should ensure—

    a. There always are scheduled dates for specific milestones
    b. Every activity is connected to at least one predecessor and at least one successor
    c. Lead or lag time should be used in each schedule
    d. Necessary sequencing of events may be described by the activity attributes

22. A schedule performance index of less than 1.0 indicates that the—

    a. Project is running behind the monetary value of the work it planned to accomplish
    b. Earned value physically accomplished thus far is 100%
    c. Project has experienced a permanent loss of time
    d. Project may not be on schedule, but the project manager need not be concerned

23. Assume you are working to sequence the activities in your schedule. You have focused on the need for predecessors and successors for your tasks. Now, you have finished, and you realize—

    a. You have a fragnet.
    b. There can be multiple predecessors for some tasks.
    c. Several identical or nearly identical series of activities are repeated throughout the project.
    d. There are too many successors.

24. To meet regulatory requirements, you need to crash your project schedule. Your first step is to compute—

    a. The cost and time slope for each critical activity that can be expedited
    b. The cost of additional resources to be added to the project's critical path
    c. The time saved in the overall schedule when tasks are expedited on the critical path
    d. Three probabilistic time estimates of PERT for each critical path activity

25. A key input to the Define Activities process is the—

    a. Project management plan
    b. Project scope statement
    c. Project scope baseline
    d. Project charter

26. Unlike bar charts, milestone charts show—

    a. Scheduled start or completion of major deliverables and key external interfaces
    b. Activity start and end dates of critical tasks
    c. Expected durations of the critical path
    d. Dependencies between complementary projects

27. Project managers should pay attention to critical and subcritical activities when evaluating project time performance. One way to do this is to analyze 10 subcritical paths in order of ascending float. This approach is part of—

    a. Variance analysis
    b. Simulation
    c. Earned value management
    d. Trend analysis

28. An activity has an early start date of the 10th and a late start date of the 19th. The activity has a duration of four days. There are no non-workdays. From the information given, what can be concluded about the activity?

    a. Total float for the activity is nine days
    b. The early finish date of the activity is the end of the day on the 14th.
    c. The late finish date is the 25th.
    d. The activity can be completed in two days if the resources devoted to it are doubled.

29. In project development, schedule information, such as who will perform the work, where the work will be performed, activity type, and WBS classification, are examples of—

    a. Activity attributes
    b. Constraints
    c. Data in the WBS repository
    d. Refinements

30. There are purposes and benefits in the Control Schedule process. Its benefit is—

    a. It maintains the schedule baseline
    b. It minimizes risk
    c. It reprioritizes remaining work as required
    d. It describes how the schedule will be managed and controlled

31. It is important to use the critical path method in Control Schedule because—

    a. It assists in reviewing scenarios to bring the schedule in line with the plan
    b. It enables a consideration of the resource availability and the project time
    c. It examines project performance over time
    d. It can help identify schedule risks

32. Several types of float are found in project networks. Float that is used by a particular activity and does NOT affect the float in later activities is called—

    a. Extra float
    b. Free float
    c. Total float
    d. Expected float

33. Assume your organization has decided to use agile for most of it projects. You are managing a project using agile and are working to develop your schedule. In doing so, you are using agile release planning. Your schedule is based on the—

    a. Resources available
    b. The requirements plan
    c. Start and finish dates from activities
    d. Product roadmap

34. You are managing a new technology project designed to improve the removal of hazardous waste from your city. You are in the planning phase of this project and have prepared your network diagram. Your next step is to—

    a. Describe any unusual sequencing in the network
    b. State the number resources required to complete each activity
    c. Establish a project calendar and link it to individual resource calendars
    d. Determine which schedule compression technique is the most appropriate, because your customer requests that the project be completed as soon as possible

35. Assume you are working to Estimate Activity Durations. You need to determine the amount of work to complete each activity and the resources required. As you work to do so, you should consider—

    a. Consulting with people who may have worked on similar projects
    b. The requirements documentation
    c. Staff motivation
    d. Adding buffers for needed but often scare resources

36. You are managing a project that will use a virtual team with team members on three different continents. Your company is looking to use the virtual team to provide a lower cost product by using resources in countries that have a favorable exchange rate to that of your country. To assist in this process as you estimate resource requirements, it is helpful to consider—

    a. Bottom-up estimating
    b. Reserve analysis
    c. Analogous estimating
    d. Assumptions analysis

37. Activity A has a pessimistic *(P)* estimate of 36 days, a most likely *(ML)* estimate of 21 days, and an optimistic *(O)* estimate of 6 days. What is the probability that activity A will be completed in 16 to 26 days?

    a. 55.70 percent
    b. 68.26 percent
    c. 95.46 percent
    d. 99.73 percent

38. You are managing a project to redesign a retail store layout to improve customer throughput and efficiency. Much project work must be done on site and will require the active participation of store employees who are life-long members of a powerful union with a reputation for labor unrest. One important component of your schedule must be—

    a. A resource capabilities matrix
    b. Buffers and reserves
    c. A resource calendar
    d. A resource histogram

39. Assume you are working on a on a project using lean manufacturing. Now you are preparing your schedule. You feel your tasks may be relatively similar in size and scope. You should consider—

    a. Fixed number of work periods
    b. Percent of the estimated activity duration
    c. Buffers
    d. Kanban system

40. The reason that the schedule performance index (SPI) is shown as a ratio is to—

    a. Enable a detailed analysis of the schedule regardless of the value of the schedule variance
    b. Distinguish between critical path and noncritical path work packages
    c. Provide the ability to show performance for a specified time period for trend analysis
    d. Measure the actual time to complete the project

Answer Sheet

1.	a	b	c	d		21.	a	b	c	d
2.	a	b	c	d		22.	a	b	c	d
3.	a	b	c	d		23.	a	b	c	d
4.	a	b	c	d		24.	a	b	c	d
5.	a	b	c	d		25.	a	b	c	d
6.	a	b	c	d		26.	a	b	c	d
7.	a	b	c	d		27.	a	b	c	d
8.	a	b	c	d		28.	a	b	c	d
9.	a	b	c	d		29.	a	b	c	d
10.	a	b	c	d		30.	a	b	c	d
11.	a	b	c	d		31.	a	b	c	d
12.	a	b	c	d		32.	a	b	c	d
13.	a	b	c	d		33.	a	b	c	d
14.	a	b	c	d		34.	a	b	c	d
15.	a	b	c	d		35.	a	b	c	d
16.	a	b	c	d		36.	a	b	c	d
17.	a	b	c	d		37.	a	b	c	d
18.	a	b	c	d		38.	a	b	c	d
19.	a	b	c	d		39.	a	b	c	d
20.	a	b	c	d		40.	a	b	c	d

Answer Key

1. d.   15

The total duration for the path B-C-D-E-I is 15. The duration of any other path in the network is less than 15. You calculate the critical path in this question through the critical path method. You are doing so to estimate the minimum project duration and the amount of schedule flexibility in the network paths. To do so, determine the early start, early finish, late start, and late finish for all the activities by performing a forward and backward pass. The critical path represents the longest path in the network, which determines the shortest possible project duration. The early and late start and finish dates are not the schedule but show the time period when the activity could be executed. The critical path method calculates the amount of scheduling flexibility on logical network paths. The critical path method critical path normally is one that has zero total float on the critical path; note the word "normally". Activities on the critical path then are critical path activities.

In the PMBOK, recognize the way PMI® has elected to show the critical path. PMI adds a day to the forward path as in Task B, it starts with a six and should be a five; the same is the case with Task C. It should start with a 5. The same is how the backward path is displayed. If you develop schedules as has just been described, it is the correct way to do so and is shown in the examples in these questions. However, for the purpose of the PMP exam, recognize the PMBOK approach may be used. [Planning]

PMI®, PMBOK® Guide, 2017, 210–211
PMI® *PMP Examination Content Outline,* 2015, Planning, 6, Task 4

2. c. 1

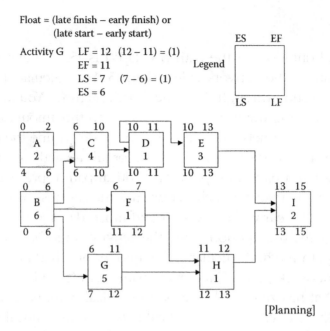

[Planning]

Float is often called slack. The critical path activities have zero float. Positive total float is caused if the backward path is calculated from a schedule constraint that is later than the early finish date calculated by the forward path. Negative total float is caused if a constraint on the late date is violated by duration and logic. Once the total float for a network path has been calculated then the free float, or the amount of time an activity can be delayed without delaying the early start of a successor or violating a schedule constraint, can be determined.

PMI®, PMBOK® Guide, 2017, 210–211

PMI® *PMP Examination Content Outline*, 2015, Planning, 5, Task 4

3. a. −1

The imposed finish date becomes the late finish for Activity I. The late dates for each activity need to be recalculated. The dates for Activity E become—

ES = 10
EF = 13
LS = 9
LF = 12

Total float = LS − ES or 9 − 10 = (−1) or
LS − EF or 12 − 13 = (−1)

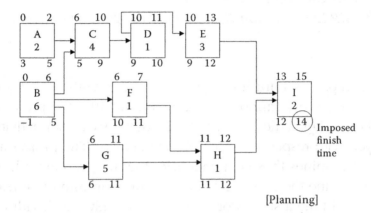

[Planning]

PMI®, PMBOK® Guide, 2017, 210–211
PMI® *PMP Examination Content Outline*, 2015, Planning, 6, Task 4

4. b. 11

The late dates for all activities need to be recalculated given the changed duration. Activity G's revised late dates are

LF = 11
LS = 6

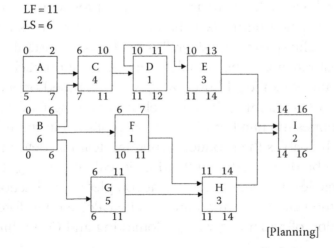

[Planning]

PMI®, *PMBOK® Guide*, 2017, 210–211
PMI® *PMP Examination Content Outline*, 2015, Planning, 6, Task 4

5. b. A milestone

   A milestone is a significant point or event in the project. Milestones may be required by the project sponsor, customer, or other external factors, such as those required by contract, for the completion of certain deliverables. They are similar to schedule activities, with the same structure and objectives, but they have zero duration as they represent a moment in time. A milestone list is an output of Define Activities. [Planning]

   PMI®, *PMBOK® Guide*, 2017, 186
   PMI® *PMP Examination Content Outline*, 2015, Planning, 6, Task 4

6. a. Lag

   For example, in a finish-to-start dependency with a 20-day lag, the successor activity cannot start until 20 days after the predecessor has finished. A lag is the amount of time that a successor activity will be delayed with respect to a predecessor activity. The project management team determines those dependencies that may require a lead or a lag to best define the logical relationship. Keep in mind that leads [the amount of time a successor activity can be advanced with respect to a predecessor activity] and lags should not replace schedule logic, and their use should be documented. [Planning]

   PMI®, *PMBOK® Guide*, 2017, 192–193
   PMI® *PMP Examination Content Outline*, 2015, Planning, 6, Task 4

7. a. Iteration burndown charts

   In Control Schedule, there are a number of tools and techniques you can use. Iteration burndown charts are an approach as part of the data analysis techniques to consider. These charts are useful to track the work that remains to be completed in the iteration backlog to evaluate the variance to an ideal burndown, which is based on work remaining from iteration planning. In agile, release planning may be done, and it determines the number of iterations or sprints in the release. This approach enables the product owner and team to determine how much needs to be developed, and the time it will take. Using the burndown chart enables the likely variances at completion and actions to take during the iteration. A trend line can be calculated to forecast the completion of remaining work. [Monitoring and Controlling]

   PMI®, *PMBOK ® Guide*, 2017, 216, 226
   PMI® *PMP Examination Content Outline*, 2015, Planning, 6, Task 4

8. a. Finish-to-start

    While many schedules use all four approaches, especially on complex projects, the most common approach is finish to start. In it a successor activity cannot start until a predecessor activity has finished. The start-to-finish approach is rarely used, but you should recognize it could be used in case you have an exam question about it. [Planning]

    PMI®, *PMBOK ® Guide*, 2017, 190
    PMI® *PMP Examination Content Outline,* 2015, Planning, 6, Task 4

9. b. Schedule management plan

    The schedule management plan is part of the overall project management plan. Whether it is formal or informal, highly detailed or broadly framed, it is based on specific project needs. It is the output of the Plan Schedule Management process. It Includes appropriate control thresholds. It establishes criteria and activities to develop, maintain, and control the schedule. [Planning]

    PMI®, *PMBOK® Guide*, 2017, 151
    PMI® *PMP Examination Content Outline,* 2015, Planning, 6, Task 4

10. b. Monte Carlo analysis

    Simulation is a data analysis tool and technique in Develop Schedule. It involves calculating multiple work package durations with different sets of assumption, constraints, risks, issues, or scenario using probability distributions and other representations of uncertainty. It shows a probability distribution for the project of achieving a certain target date or project finish date. [Planning]

    PMI®, *PMBOK® Guide*, 2017, 213
    PMI® *PMP Examination Content Outline,* 2015, Planning, 6, Task 4

11. b. 70 weeks

$$E(t) = \frac{\text{Optimistic} + (4 \times \text{Most likely}) + \text{Pessimistic}}{6}$$
$$= \frac{40 + 200 + 180}{6} = \frac{420}{6} = 70 \text{ weeks}$$

Use of the three-point estimates [optimistic, most likely, and pessimistic] is a tool and technique used in Estimate Activity Duration. The optimistic estimate is based on the duration of the activity given the resources and their productivity, realistic estimates of their availability for the activity, dependencies on other participants, and interruptions. The optimistic estimate is the duration based on analysis of the best-case scenario for the activity. The pessimistic estimate in contrast is the duration based on the worst-case scenario for the activity. Depending on the distribution of values in the range of the three estimates, the expected duration can be calculated. A commonly-used formula is the triangular distribution especially if there is insufficient historical information or judgmental data. Using three-point estimates with an assumed distribution provide an expected duration and clarify the range of uncertainty around the expected duration. [Planning]

PMI®, *PMBOK® Guide*, 2017, 201
PMI® *PMP Examination Content Outline*, 2015, Planning, 6, Task 4

12. b. Fast tracking

Fast tracking is a way to accelerate the project schedule. It is a schedule compression tool and technique in the Develop Schedule process in which activities or phases normally done in sequence are performed in parallel for part of their duration. However, it may result in rework and increased risk. Consider it for activities that can be overlapped to shorten the project schedule. It only works effectively when activities can be overlapped to shorten the project duration on the critical path. Using leads in case of schedule acceleration may increase coordination efforts between the activities concerned. It may increase the project's costs and risks to quality. [Planning]

PMI®, *PMBOK® Guide*, 2017, 215
PMI® *PMP Examination Content Outline*, 2015, Planning, 6, Task 4

13. b. Calendar time between the start of A to the finish of B is 11 days

The duration of A, which is three, is added to the duration of B, which is four, for a total of seven. The three days between the activities is lag and not duration. The lag is a constraint and must be taken into account as part of the network calculations, but it does not consume resources. The total time by the calendar is 11 days as counted from the morning of Monday the 4th. The lag occurs over Thursday, Friday, and Saturday. Sunday is a non-work day, so activity B does not start until Monday the 11th. Therefore, the calendar time is 11 days, and activity B ends on Thursday the 14th. In the Sequence Activities and Develop Schedule processes, leads and lags are a tool and technique [Planning]

PMI®, *PMBOK® Guide*, 2017, 192–193, 214
PMI® *PMP Examination Content Outline*, 2015, Planning, 6, Task 4

14. b. Analogous estimating

Analogous estimating is a tool and technique in Estimate Activity Durations. Although limitations exist with all estimating approach, analogous estimating is often used when there is a limited amount of information for the project. It uses historical information from a similar activity or a project. It uses data from a previous but similar project such as duration, budget, size, weight, and complexity to estimate the same data or measure for a future project. It is considered as a gross value estimating approach, which often is adjusted for known differences in project complexity. [Planning]

PMI®, P*MBOK® Guide*, 2017, 200
PMI® *PMP Examination Content Outline*, 2015, Planning, 6, Task 4

15. d. Mandatory or hard

Mandatory dependencies may be required contractually or may be inherent in the nature of the project work. Dependency determination and integration is a tool and technique in the Sequence Activity process. They describe a relationship in which the successor activity cannot be started because of physical constraints until the predecessor activity has been finished. For example, software cannot be tested until it has been developed (or coded). They may be called hard logic or hard dependencies; however, technical dependencies may not be mandatory. This distinction is important to remember. The project team determines the dependencies that are mandatory as it sequences the activities. However, these dependencies should not be confused with assigning constraints in the scheduling tool. [Planning]

PMI®, *PMBOK® Guide*, 2017, 191
PMI® *PMP Examination Content Outline*, 2015, Planning, 6, Task 4

16. c. Schedule baseline

    The schedule baseline is a key input to the Control Schedule process. It is compared with actual results to determine if a change requires corrective or preventive action. [Monitoring and Controlling]

    PMI®, *PMBOK® Guide*, 2017, 224
    PMI® *PMP Examination Content Outline*, 2015, Monitoring and Controlling, 9, Task 2

17. d. Resource leveling

    While resource leveling will often result in a project duration that is longer than the preliminary schedule as the original critical path probably will change and increase, it can also be used to get a schedule back on track by reassigning activities from noncritical to critical path activities. It is a resource optimization technique in the Develop Schedule process. Start and finish dates are adjusted based on resource constraints, and the goal is to balance demand with the available supply of resources. It is used when scarce or critically required resources are only available at certain times or in limited quantities, are over allocated if the resource is assigned to two activities at the same time, or to keep resource use at a constant level. [Planning]

    PMI®, *PMBOK® Guide*, 2017, 211
    PMI® *PMP Examination Content Outline*, 2015, Planning, 6, Task 4

18. d. C, A, F, and G

    First, it is necessary to determine the critical path, which is A, C, F, and G. To determine the lowest weekly crashing cost, start with C at $1,500 per week. The next activity is A, followed by F and G. Understand how to crash a schedule in case you have a question on it in your exam. It is a schedule compression technique in the Develop Schedule process. The purpose is to shorten the schedule for the least incremental cost by adding resources. Examples are to approve overtime, add additional resources, or pay to expedite delivery of activities on the critical path. It only works for activities on the critical path and does not always provide a viable alternative. The latter means it can result in increased risk and/or cost. [Planning]

    PMI®, *PMBOK® Guide*, 2017, 181
    PMI® *PMP Examination Content Outline*, 2015, Planning, 6, Task 4

19. b. Discretionary dependency

This situation is an example of a discretionary dependency. It is a dependency determination and integration tool and technique in the Sequence Activity process. These dependencies may be called preferred logic or soft logic They are established based on knowledge of best practices in an application area or in an unusual aspect of the project where a specific sequence may be desired. They often are used in construction, and the order is not mandatory, but if the activities are performed in a sequential order, the overall risk is reduced. As a project manager, you should document these dependencies as they can create arbitrary total float and can limit later scheduling options. The project team makes the decision as to which dependencies are discretionary. [Planning]

PMI®, *PMBOK® Guide*, 2017, 191
PMI® *PMP Examination Content Outline*, 2015, Planning, 6, Task 4

20. c. The final output is described as activities

In the Create WBS process, the work package is defined as the lowest level of the WBS to then estimate and manage cost and duration. In the Define Activities process, the final output is described as activities. It is a tool and technique in both processes as it divides and subdivides the project scope and deliverables into smaller, more manageable parts. Activities represent the effort to create the work package in Create WBS. The best approach is to involve team members in decomposition in both processes. [Planning]

PMI®, *PMBOK® Guide*, 2017, 185
PMI® *PMP Examination Content Outline*, 2015, Planning, 6, Task 4

21. b. Every activity is connected to at least one predecessor and at least one successor

The Sequence Activity process involves identifying and documenting relationships among the project activities. Every activity is connected to at least one predecessor or successor except the first and last. Logical relationships are set up to create a realistic project schedule. Some leads and lags may be used. The purpose of the Sequence Activity process is to convert the activity list into a diagram, which is the first step in establishing the schedule baseline. [Planning]

PMI®, *PMBOK® Guide*, 2017, 188
PMI® *PMP Examination Content Outline*, 2015, Planning, 6, Task 4

22. a. Project is running behind the monetary value of the work it planned to accomplish

    The SPI represents how much of the originally scheduled work has been accomplished at a given period in time, thus providing the project team with insight as to whether the project is on schedule. It is a measure of schedule efficiency and is calculated by EV/PV. A SPI of 1 means the project is on schedule, and the work that has been done to date is exactly the same as the work that was planned. Other values show how much costs are over or under [as in this example] the budgeted amount for the work planned. [Monitoring and Controlling]

    PMI®, *PMBOK® Guide*, 2017, 226
    PMI® *PMP Examination Content Outline,* 2015, Monitoring and Controlling, 9, Task 1

23. b. There can be multiple predecessors for some tasks.

    The main output in Sequence Activities is a project network diagram of the logical relationships or dependencies in the schedule activities. A summary narrative can accompany the diagram. You also may have multiple predecessors for a task or activity, which is known as path convergence. Activities with multiple successors are known as path divergence. In both of these situations, these activities or tasks are at greater risk. [Planning]

    PMI®, *PMBOK® Guide*, 2017, 194
    PMI® *PMP Examination Content Outline,* 2015, Planning, 6, Task 4

24. a. The cost and time slope for each critical activity that can be expedited

    Slope = (Crash cost – Normal cost)/(Crash time – Normal time). This calculation shows the cost per day of crashing the project. The slope is negative to indicate that as the time required for a project or task decreases, the cost increases. If the costs and times are the same regardless of whether they are crashed or normal, the activity cannot be expedited. Crashing only works well for those activities on the critical path, may not produce a viable alternative, and may result in increased risk or cost. It is important to know how it is done in case you might see an exam question about how to do it. [Planning]

    PMI®, *PMBOK® Guide*, 2017, 215
    PMI® *PMP Examination Content Outline,* 2015, Planning, 6, Task 4

25. a. Project management plan

    The project management plan is the key input to Define Activities. It includes the project's components, and it also includes the schedule management plan and the scope baseline. The schedule management plan defines the schedule methodology; waves if rolling wave planning is done; and the necessary level of detail for the project. The scope baseline shows the WBS, scope statement, and WBS dictionary. It also contains deliverables, constraints, and assumptions as documented in the scope statement. [Planning]

    PMI®, *PMBOK® Guide*, 2017, 184
    PMI® *PMP Examination Content Outline,* 2015, Planning, 6, Task 4

26. a. Scheduled start or completion of major deliverables and key external interfaces

    Milestones are singular points in time, such as the start or completion of a significant activity or group of activities. Milestone charts are an output of the Develop Schedule process, and while they are similar to bar charts, their focus is only on the scheduled start or completion of major deliverables and key external influences. [Planning]

    PMI®, *PMBOK® Guide*, 2017, 218
    PMI® *PMP Examination Content Outline,* 2015, Planning, 6, Task 4

27. a. Variance analysis

    Performance of variance analysis during Control Schedule is a tool and technique that can be used. It focuses on variance analysis in planned versus actual start and finish dates, planned versus actual durations, and variances in float. Float variance is an essential planning component for evaluating project time performance. It determines the cause and degree of variance relative to the schedule baseline, estimates the implications of the variances for future work to complete the project, and helps decide if corrective or preventive action is needed. It may not be needed on some activities not on the critical path, but an activity on the critical path may need immediate attention. [Monitoring and Controlling]

    PMI®, *PMBOK® Guide*, 2017, 227
    PMI® *PMP Examination Content Outline,* 2015, Monitoring and Controlling 9, Task 1

28. a. Total float for the activity is nine days.

    Total float or slack is computed by subtracting the early start date from the late start date, or 19 − 10 = 9. To compute the early finish date given a duration of 4, we would start counting the activity on the morning of the 10th; therefore, the activity would be completed at the end of day 13, not 14 (10, 11, 12, 13). If we started the activity on its late start date on the morning of the 19th, we would finish at the end of day 22, not 25. Insufficient information is provided to determine whether this activity can be completed in 2 days if the resources are doubled. [Planning]

    PMI®, *PMBOK® Guide*, 2017, 210–211
    PMI® *PMP Examination Content Outline*, 2015, Planning, 6, Task 4

29. a. Activity attributes

    Identifying activity attributes is helpful for further selection and sorting of planned activities. They are used for schedule development, for report formatting purposes, and for selecting, ordering, and sorting planned schedule activities in various ways in reports. They extend the definition of activities by identifying multiple components associated with each activity. They include an activity identifier that is unique, the WBS identification number, and activity label or name. Once they are finished, they may include activity predecessors and successors, logical relationships, leads and lags, resource requirements, imposed dates, constraints and assumptions. They can be used to determine where the work has to be performed, the project calendar, the person responsible, and the type of activity involved. [Planning]

    PMI®, *PMBOK® Guide*, 2017, 186
    PMI® *PMP Examination Content Outline*, 2015, Planning, 6, Task 4

30. a. It maintains the schedule baseline

    In Control Schedule the benefit is it maintains the schedule baseline throughout the project. The purpose of the Control Schedule process is to update the schedule and manage any changes to the baseline. [Monitoring and Controlling]

    PMI®, *PMBOK® Guide*, 2017, 222
    PMI® *PMP Examination Content Outline*, 2015, Monitoring and Controlling, 9, Task 4

31. d. It can help identify schedule risks

    In Control Schedule, the critical path method is a tool and technique to use. It compares progress along the critical path to help determine schedule status. Any variance on the critical path has an impact on the project's end date. Also, evaluating progress of the critical path activities on a near critical path can help identify schedule risk. [Monitoring and Controlling]

    PMI®, *PMBOK® Guide*, 2017, 227
    PMI® *PMP Examination Content Outline*, 2015, Monitoring and Controlling, 9, Task 1

32. b. Free float

    Free float is defined as the amount of time an activity can be delayed without delaying the early start of any immediately succeeding activities or violating a schedule constraint. It is important to know the difference between zero total float, positive total float, negative total float, and free float and be able to recognize examples of all four types of float. [Planning]

    PMI®, *PMBOK® Guide*, 2017, 210
    PMI® *PMP Examination Content Outline*, 2015, Planning, 6, Task 4

33. d. Product roadmap

    When using agile release planning, it provides a high-level summary timeline of the release schedule, which tends to be three to six months. It is based on the product roadmap and the product vison for the project's execution. It also determines the number of iterations or sprints in the release and enables the product owner and team to decide how much needs to be developed and how long it will take to have a reasonable product based on business goals, dependencies, and any impediments. [Planning]

    PMI®, *PMBOK® Guide*, 2017, 216
    PMI® *PMP Examination Content Outline*, 2015, Planning, 6, Task 4

34. a. Describe any unusual sequencing in the network

A summary narrative can accompany the schedule network diagram and describe the basic approach used to sequence the activities in the network. This narrative also should describe any unusual sequences in the network. The project schedule network diagrams are an output of the Sequence Activities process. The project schedule network diagram is a graphic representation of the logical relationships or dependencies among the project schedule activities. [Planning]

PMI®, *PMBOK® Guide*, 2017, 194
PMI® *PMP Examination Content Outline*, 2015, Planning, 6, Task 4

35. c. Staff motivation

The Estimate Activity durations process determines the number of work periods needed to complete the activity using the resource calendars. It is difficult to do especially considering the nature of the work involved and any constraints. You also need to consider staff motivation as the project manager must be aware of the Student Syndrome, which is procrastination. Many people, especially if they are not working full time on the project, may tend to apply themselves at the last minute possible before the deadline. Parkinson's Law also may apply where work expands to fill the time available for completion. [Planning]

PMI®, *PMBOK® Guide*, 2017, 197
PMI® *PMP Examination Content Outline*, 2015, Planning, 6, Task 4

36. b. Reserve analysis

In estimating activity resources, reserve analysis is a data analysis tool and technique. In it, the amount of contingency and management reserves required are determined, which fits the situation in the question. Contingency reserves may be called schedule reserves for schedule uncertainty. Contingency reserves are the estimated duration within the schedule baseline used for any schedule-related risks. They may be associated with the percent of the estimated duration or a fixed number of work periods. Any use of contingency reserves should be documented as should management reserves. Management reserves are a specified amount of the project budget withheld for management control purposes. They are used for any unknown-unknowns. [Planning]

PMI®, *PMBOK® Guide*, 2017, 202
PMI® *PMP Examination Content Outline*, 2015, Planning, 6, Task 4

37. b. 68.26 percent

First, compute the standard deviation:

$$\sigma = \frac{P - O}{6} \quad \text{or} \quad \frac{36 - 6}{6} = 5 \text{ days}$$

Next, compute three-point expected time:

$$\frac{P + 4(ML) + O}{6} \quad \text{or} \quad \frac{36 + 4(21) + 6}{6} = 21 \text{ days}$$

Finally, determine range of outcomes using 1σ:

$21 - 5 = 16$ days, and $21 + 5 = 26$ days

Simply defined, 1σ is the amount on either side of the mean of a normal distribution that will contain approximately 68.26 percent of the population. Duration estimates based on three-point estimates with an assumed distribution provide an expected duration and clarify the range of uncertainty around the expected duration. [Planning]

PMI®, *PMBOK® Guide*, 2017, 201
PMI® *PMP Examination Content Outline*, 2015, Planning, 6, Task 4

38. c. A resource calendar

Project and resource calendars identify periods when work is allowed. Project calendars affect all resources. Resource calendars affect a specific resource or a resource category, such as a labor contract that requires certain workers to work on certain days of the week. The resource calendar identifies the working days and shifts that each specific resource is available and is used to estimate resource use. The resource calendars specify when and how long identified project resources will be available during the project, and this information may be at the activity or project level. This knowledge includes considerations of resource experience and/or skill level and where the resources are located and when they may be available. Resource calendars are an input to Estimate Activity Durations. [Planning]

PMI®, *PMBOK® Guide*, 2017, 199
PMI® *PMP Examination Content Outline*, 2015, Planning, 6, Task 4

39. d. Kanban system

    The Kanban system is also known as on-demand scheduling. It is based on the theory of constraints and pull-based scheduling concepts from lean manufacturing as described in the question. The idea is to limit the team's work in progress to balance demand against the team's delivery throughout. It does not rely on a previously developed schedule but instead pulls work from a backlog or intermediate queue of work to be done immediately when resources become available. It often is used for projects that evolve from the product requirements incrementally or as in the question where takes may be similar in size and scope. [Planning]

    PMI®, *PMBOK® Guide*, 2017, 177
    PMI® *PMP Examination Content Outline*, 2015, Planning, 6, Task 4

40. c. Provide the ability to show performance for a specified time period for trend analysis

    Because schedule performance index (SPI) and cost performance index (CPI) are expressed as ratios, they can be used to show performance for a specific time period or trends over a long-time horizon. The SPI measures schedule efficiency as EV/PV and measures how efficiently the project team is using its time. If it is less than 1, less work was completed than planned; greater than 1 shows more work was completed than was planned. It is important to also analyze performance on the critical path to determine if the project will finish ahead of or behind finish dates. [Monitoring and Controlling]

    PMI®, *PMBOK® Guide*, 2017, 226
    PMI® *PMP Examination Content Outline*, 2015, Monitoring and Controlling, 9, Task 1

Project Cost Management

Practice Questions

INSTRUCTIONS: Note the most suitable answer for each multiple-choice question in the appropriate space on the answer sheet.

You are using earned value progress reporting for your current project in an effort to teach your software developers the benefits of earned value. You plan to display project results on the PMO dashboard so that the team knows how the project is progressing. Use the current status, listed below, to answer questions 1 through 4:

PV = $2,200
EV = $2,000
AC = $2,500
BAC = $10,000

1. According to earned value analysis, the SV and status of the project described above is—

   a. −$300; the project is ahead of schedule
   b. +$8,000; the project is on schedule
   c. +$200; the project is ahead of schedule
   d. −$200; the project is behind schedule

2. What is the CPI for this project, and what does it tell us about cost performance thus far?

   a. 0.20; actual costs are exactly as planned
   b. 0.80; actual costs have exceeded planned costs
   c. 0.80; actual costs are less than planned costs
   d. 1.25; actual costs have exceeded planned costs

3. The CV for this project is—

   a. 300
   b. −$300
   c. 500
   d. −$500

4. What is the EAC for this project, and what does it represent?

   a. $12,500; the revised estimate for total project cost (based on performance thus far)
   b. $10,000; the revised estimate for total project cost (based on performance thus far)
   c. $12,500; the original project budget
   d. $10,000; the original project budget

5. You have now prepared your cost management plan so now you are preparing your project's cost estimate. You decided to use analogous estimating. Analogous estimating—

   a. Supports top-down estimating
   b. Calculates costs based on project parameters
   c. Produces higher levels of accuracy
   d. Indicates a range of estimates

6. While cost estimates are the main output of the Estimate Costs process, not to be overlooked is—

   a. Updates to the cost management plan
   b. Assumptions log
   c. Updates to the project schedule
   d. Cost baseline

7. You must consider direct costs, indirect costs, overhead costs, and general and administrative costs during cost estimating. An example of an indirect cost is—

   a. Salary of the project manager
   b. Subcontractor expenses
   c. Materials used by the project
   d. Electricity

8. If the cost variance is the same as the schedule variance and both numbers are greater than zero, then—

   a. The cost variance is due to the schedule variance
   b. The variance is favorable to the project
   c. The schedule variance can be easily corrected
   d. Labor rates have escalated since the project began

9. You are responsible for preparing a cost estimate for a large World Bank project. You decide to prepare a bottom-up estimate because your estimate needs to be as accurate as possible. Your first step is to—

   a. Locate a computerized tool to assist in the process
   b. Use the cost estimate from a previous project to help you prepare this estimate
   c. Identify and estimate the cost for each work package or activity
   d. Consult with subject matter experts and use their suggestions as the basis for your estimate

10. Management has grown weary of the many surprises, mostly negative, that occur on your projects. In an effort to provide stakeholders with an effective performance metric, you will use the to-complete performance index (TCPI). Its purpose is to—

    a. Determine the schedule and cost performance needed to complete the remaining work within management's financial goal for the project
    b. Determine the cost performance needed to complete the remaining resources to meet management's financial goal
    c. Predict final project costs
    d. Predict final project schedule and costs

11. If operations on a work package were estimated to cost $1,500 and finish today but, instead, have cost $1,350 and are only two-thirds complete, the cost variance is—

    a. $150
    b. –$150
    c. –$350
    d. –$500

12. When you review cost performance data on your project, different responses will be required depending on the degree of variance or control thresholds from the baseline. For example, a variance of 10 percent might not require immediate action, whereas a variance of 100 percent will require investigation. A description of how you plan to manage cost variances should be included in the—

    a. Cost management plan
    b. Change management plan
    c. Performance measurement plan
    d. Variance management plan

13. As of the fourth month on the Acme project, cumulative planned expenditures were $100,000. Actual expenditures totaled $120,000. How is the Acme project doing?

    a. It is ahead of schedule.
    b. It is in trouble because of a cost overrun.
    c. It will finish within the original budget.
    d. The information is insufficient to make an assessment.

14. On your project, you need to assign costs to the time period in which they are incurred. To do this, you should—

    a. Identify the project components so that costs can be allocated
    b. Use the project schedule as an input to determine the budget
    c. Prepare a detailed and accurate cost estimate
    d. Prepare a cost performance plan

15. You have a number of costs to track and manage because your project is technically very complex. They include direct costs and indirect (overhead) costs. You have found that managing overhead costs is particularly difficult because they—

    a. Are handled on a project-by-project basis
    b. Represent only direct labor costs
    c. Represent only equipment and materials needed for the project
    d. Can be included in the activities or at higher levels

16. If you want to calculate the ETC based on the assumption that work is proceeding as planned, the remaining work can be calculated by which of the following formulas?

    a. $ETC = BAC - EV$
    b. $ETC = (BAC - EV)/CPI$
    c. $ETC = AC + EAC$
    d. $ETC = EAC - AC$

17. You receive a frantic phone call from your vice president who says she is going to meet with a prospective client in 15 minutes to discuss a large and complex project. She asks you how much the project will cost. You quickly think of some similar past projects, factor in a few unknowns, and give her a number. You provided which of the following type of estimate—

    a. Definitive
    b. Budget
    c. Rough-order-of-magnitude
    d. Detailed

18. Your approved cost baseline has changed because of a major scope change on your project. Your next step should be to—

    a. Estimate the magnitude of the scope change
    b. Issue a change request
    c. Document lessons learned
    d. Execute the approved scope change

19. You have set aside a certain portion of your project for contingency and management reserves. You want to use these reserves to cover the costs of risk responses or other contingencies. On your project, you have had some opportunities that resulted, which means you should—

    a. Add these opportunities to your reserves for later risk responses
    b. Use value analysis
    c. Remove them from the contingency fund but maintain the management reserves
    d. Use value engineering

20. There are many useful EVM metrics, but the most critical is—

    a. CPI
    b. EAC
    c. TCPI
    d. VAC

21. The approved, integrated scope-schedule-cost plan for project work is—

    a. Project budget baseline
    b. Performance measurement baseline
    c. Level-of-effort control accounts
    d. Expressed in the To Complete Performance Index

22. It is expensive to lease office space in cities around the world. Office space can cost approximately USD $150 per square foot in Tampa, Florida. And it can cost approximately ¥50,000 per square meter in Tokyo. These "averages" can help a person to determine how much it will cost to lease office space in these cities based on the amount of space leased. These estimates are examples of—

    a. Variance analysis
    b. Parametric estimating
    c. Bottom-up estimating
    d. Reserve analysis

23. Your project manager has requested that you provide him with a forecast of project costs for the next 12 months. He needs this information to determine if the budget should be increased or decreased on this major construction project. In addition to the usual information sources, which of the following should you also consider?

    a. Cost estimates from similar projects
    b. WBS
    c. Project schedule
    d. Costs that have been authorized and incurred

24. There are a number of different earned value management rules of performance measurement that can be established as part of the cost management plan. One of them is the—

    a. Control account
    b. Formulas to determine the ETC
    c. Risk thresholds
    d. Activity attributes

25. Which of the following calculations CANNOT be used to determine EAC?

    a. $EAC = ETC - AC$
    b. $EAC = BAC/CPI$
    c. $EAC = AC + (BAC - EV)$
    d. $EAC = AC + (BAC - EV)/(CPI \times SPI)$

26. Typically, the statement "no one likes to estimate, because they know their estimate will be proven incorrect" is true. However, you have been given the challenge of estimating the costs for your nuclear reactor project. You are considering a variety of items as you incorporate the cost of financing, which is in the—

    a. Schedule
    b. Financial management plan
    c. The amount of indirect costs
    d. The benefit delivery plan

27. By reviewing cumulative cost curves, the project manager can monitor—

    a. EAC
    b. PV, EV, and AC
    c. CVs and SVs
    d. CPI and SPI

28. Control accounts—

    a. Are charge accounts for personnel time management
    b. Summarize project costs at level 2 of the WBS
    c. Identify and track management reserves
    d. Represent the basic level at which project performance is measured and reported

29. Assume you are holding performance review meetings on your project to assess schedule activity and work packages over-running or under-running the budget and to determine any estimated funds needed to complete work in progress. You also are using earned value. You are preparing reports for your stakeholders as they want regular cost forecasts. This means they are interested in—

    a. The Estimate at Completion
    b. CPI and SPI
    c. To Complete Performance Index
    d. Variance analysis

30. Overall cost estimates must be allocated to individual activities to establish the cost performance baseline. In an ideal situation, a project manager would prefer to prepare estimates—

    a. Before the budget is complete
    b. After the budget is approved by management
    c. Using a parametric estimating technique and model specific for that project type
    d. Using a bottom-up estimating technique

31. One way to engage team members to improve estimate accuracy is to—

    a. Hold a focus group
    b. Use vendor bid analysis
    c. Use decision making
    d. Use a Delphi technique

32. Assume you have used reserve analysis as a tool and technique in Control Costs. However, during your project risk identification and subsequent analysis of the identified risks was ongoing. In fact you had a risk expert on your team. This means you may need to—

    a. Use management reserve
    b. Rebaseline your cost estimate
    c. Perform an assumptions analysis
    d. Request additional contingency reserves for your budget

33. Increased attention to return on investment (ROI) now requires you to re-estimate the costs for your project. When you looked at how costs were first estimated, you realized an order-of-magnitude estimate was prepared, which was never refined. Therefore, now you are estimating costs from the beginning. As you are preparing a new smart TV along with an automated home as the products of your project, you realize market conditions should be reviewed because—

    a. A competitor may be working on the same product
    b. Global supply and demand conditions can influence resource costs
    c. Resource cost data are available in your company's knowledge transfer system
    d. Expert judgment can assist in determining when costs exceed profit

34. A revised cost baseline may be required in cost control when—

    a. CVs are severe, and a realistic measure of performance is needed
    b. Updated cost estimates are prepared and distributed to stakeholders
    c. Corrective action must be taken to bring expected future performance in line with the project plan
    d. EAC shows that additional funds are needed to complete the project even if a scope change is not needed

35. As project manager, you identified a number of acceptable tolerances as part of your earned value management system. During execution, some "unacceptable" variances occurred. After each "unacceptable" variance occurred, a best practice is to—

    a. Update the budget
    b. Prepare a revised cost estimate
    c. Adjust the project plan
    d. Document lessons learned

36. Assume that the project cost estimates have been prepared for each activity and the basis of these estimates has been determined. Now, as the project manager for your nutrition awareness program in your hospital, you are preparing your budget. Because you have estimates for more than 1,200 separate activities, you have decided to first—

    a. Aggregate these estimates by work packages
    b. Aggregate these estimates by control accounts to facilitate the use of earned value management
    c. Use the results of previous projects to predict total costs
    d. Set your cost performance baseline

37. Assume you are using earned value on your project, and as of today, your cumulative CPI is below the baseline. This means—

    a. All future work will need to be performed immediately with the range of the TCPI.
    b. The ETC work will be performed at an efficiency rate that considers both the schedule and performance indices.
    c. What the project has experienced to date is expected to continue.
    d. The budget should be rebaselined.

38. Assume it has become obvious on your project that the budget at completion is no longer viable. This means as the project manager, you should—

    a. Recommend to your managers that the project be terminated
    b. Perform an analysis of the remaining tasks on the critical path and take corrective or preventive action as required
    c. Determine the relationship of physical performance to costs spent to show performance for a specified time period for trend analysis
    d. Consider the forecasted EAC

39. Assume that your actual costs are $800; your planned value is $1,200; and your earned value is $1,000. Based on these data, what can be determined regarding your schedule variance?

    a. At +$200, the situation is favorable as physical progress is being accomplished ahead of your plan.
    b. At −$200, the physical progress is being accomplished at a slower rate than is planned, indicating an unfavorable situation.
    c. At +$400, the situation is favorable as physical progress is being accomplished at a lower cost than was forecasted.
    d. At −$200, you have a behind-schedule condition, and your critical path has slipped.

40. The CPI on your project is 0.84. This means that you should—

    a. Place emphasis on improving the timeliness of the physical progress
    b. Reassess the life-cycle costs of your product, including the length of the life-cycle phase
    c. Recognize that your original estimates were fundamentally flawed, and your project is in an atypical situation
    d. Place emphasis on improving the productivity by which work was being performed

Answer Sheet

1.	a	b	c	d		21.	a	b	c	d
2.	a	b	c	d		22.	a	b	c	d
3.	a	b	c	d		23.	a	b	c	d
4.	a	b	c	d		24.	a	b	c	d
5.	a	b	c	d		25.	a	b	c	d
6.	a	b	c	d		26.	a	b	c	d
7.	a	b	c	d		27.	a	b	c	d
8.	a	b	c	d		28.	a	b	c	d
9.	a	b	c	d		29.	a	b	c	d
10.	a	b	c	d		30.	a	b	c	d
11.	a	b	c	d		31.	a	b	c	d
12.	a	b	c	d		32.	a	b	c	d
13.	a	b	c	d		33.	a	b	c	d
14.	a	b	c	d		34.	a	b	c	d
15.	a	b	c	d		35.	a	b	c	d
16.	a	b	c	d		36.	a	b	c	d
17.	a	b	c	d		37.	a	b	c	d
18.	a	b	c	d		38.	a	b	c	d
19.	a	b	c	d		39.	a	b	c	d
20.	a	b	c	d		40.	a	b	c	d

Answer Key

1. d. –$200; the project is behind schedule

    SV is calculated as EV – PV (in this case, $2,000 – $2,200). It is the amount by which the project is ahead or behind the planned delivery date at a given point of time. If it is positive, it shows the project is ahead of schedule; if it is neutral, it is on schedule. In this example, there is a negative variance means that the work completed is less than what was planned for at that point in the project, and it is behind schedule. [Monitoring and Controlling]

    PMI®, *PMBOK® Guide*, 2017, 262
    PMI® *PMP Examination Content Outline*, 2015, Monitoring and Controlling, 9, Task 1

2. b. 0.80; actual costs have exceeded planned costs

    CPI is calculated as EV/AC (in this case, $2,000/$2,500). EV measures the budgeted dollar value of the work that has actually been accomplished, whereas AC measures the actual cost of getting that work done. If the two numbers are the same, work on the project is being accomplished for exactly the budgeted amount of money (and the ratio will be equal to 1.0). If actual costs exceed budgeted costs (as in this example), AC will be larger than EV, and the ratio will be less than 1.0. CPI is also an index of efficiency. In this example, an index of 0.80 (or 80 percent) means that for every dollar spent on the project only 80 cents worth of work is actually accomplished. At this time, the project is over the planned costs. [Monitoring and Controlling]

    PMI®, *PMBOK® Guide*, 2017, 263
    PMI® *PMP Examination Content Outline*, 2015, Monitoring and Controlling, 9, Task 1

3. d. –$500

    CV is calculated as EV – AC (in this case, $2,000 – $2,500). A negative CV means that accomplishing work on the project is costing more than was budgeted. CV measures the amount of budget surplus or deficit at a point in time, and if it is positive, it is under the planned costs; neutral means it is on the planned costs. [Monitoring and Controlling]

    PMI®, *PMBOK® Guide*, 2017, 262
    PMI® *PMP Examination Content Outline*, 2015, Monitoring and Controlling, 9, Task 1

4. a. $12,500; the revised estimate for total project cost (based on performance thus far)

EAC is calculated as BAC/CPI (in this case, $10,000/0.80). It is now known that the project will cost more than the original estimate of $10,000. The project has been getting only 80 cents worth of work done for every dollar spent (CPI), and this information has been used to forecast total project costs. This approach assumes that performance for the remainder of the project will also be based on a CPI of 0.80. The EAC is the expected total cost of completing all work expressed as the sum of the actual cost and the estimate to complete (ETC). Recognize there are three common ways to calculate the EAC based on different assumptions and learn when to use each method. [Monitoring and Controlling]

PMI®, *PMBOK® Guide*, 2017, 224
PMI® *PMP Examination Content Outline*, 2015, Monitoring and Controlling, 9, Task 1

5. a. Supports top-down estimating

A frequently used method of Estimate Costs, the analogous technique relies on experience and knowledge gained to predict future events. This technique provides planners with some idea of the magnitude of project costs but generally not within ±10%. It is considered as a gross value estimating technique, and it is generally less costly and less time consuming than other techniques but also is generally less accurate. Sometimes it is adjusted for project complexities. It is more reliable when the previous projects are similar in fact, and when the project team members doing the estimating have the needed expertise. [Planning]

PMI®, *PMBOK® Guide*, 2017, 244
PMI® *PMP Examination Content Outline*, 2015, Planning, 6, Task 3

6. b. Assumptions log

The Estimate Costs process includes updates to three project documents as an output of this process. One is the assumptions log as during this project, new assumptions may be made, new constraints may be identified, and existing constraints or assumptions may be reviewed and changed; therefore, the assumptions log is updated with this new information. Other documents to update are the lessons learned register and the risk register. [Planning]

PMI®, *PMBOK® Guide*, 2017, 247
PMI® *PMP Examination Content Outline*, 2015, Planning, 6, Task 3

7. d. Electricity

    Direct costs are incurred for the exclusive benefit of a project (for example, salary of the project manager, materials used by the project, and subcontractor expenses). Indirect costs, also called overhead costs, are allocated to a project by its performing organization as a cost of doing business. These costs cannot be traced to a specific project and are accumulated and allocated equitably over multiple projects (for example, security guards, fringe benefits, and electricity). Other examples of costs to estimate are cost of financing including any interest if needed, inflation allowance, exchange rates, or a contingency reserve. Costs are estimated for all resources to be charged to the project. [Planning]

    PMI®, *PMBOK® Guide*, 2017, 246
    PMI® *PMP Examination Content Outline,* 2015, Planning, 6, Task 3

8. b. The variance is favorable to the project

    A positive schedule variance indicates that the project is ahead of schedule. A positive cost variance indicates that the project has incurred less cost than estimated for the work accomplished; therefore, the project is under budget and ahead of schedule. The CV is calculated as EV-AC, and the SV is calculated as the EV-PV. [Monitoring and Controlling]

    PMI®, *PMBOK® Guide*, 2017, 262
    PMI® *PMP Examination Content Outline,* 2015, Monitoring and Controlling, 9, Task 1

9. c. Identify and estimate the cost for each work package or activity

    Bottom-up estimating is a method of estimating a component of work. It is derived by first estimating the cost of the individual work packages or activities to the greatest level of specified detail. Then this detailed cost is summarized or 'rolled-up' to higher levels for reporting and tracking. The cost and accuracy are influenced by the size and other attributes of the activity or the work package. It is a tool and technique in the Estimate Costs process. [Planning]

    PMI®, *PMBOK® Guide*, 2017, 244
    PMI® *PMP Examination Content Outline,* 2015, Planning, 6, Task 3

10. b. Determine the cost performance needed to complete the remaining resources to meet management's financial goal

The TCPI takes the value of work remaining and divides it by the value of funds remaining to obtain the cost performance factor needed to complete all remaining work according to a financial goal set by management. There are two methods to calculate the TCPI, and you should be familiar with each one and when it should be used. In this example, the TCPI is calculated by (BAC-EV)/(BAC-AC). If it is greater than one, it is harder to complete; if it is 1, it is the same to complete, and less than one means it is easier to complete. It thus measures the cost performance that must be achieved with the remaining resources to meet a specified management goal and is expressed as a ratio of the cost to finish the outstanding work to the available budget. If it is obvious the BAC is no longer viable, the project manager then considers the forecasted EAC. Once approved, the EAC may replace the BAC in the formula. [Monitoring and Controlling]

PMI®, *PMBOK® Guide*, 2017, 266
PMI® *PMP Examination Content Outline*, 2015, Monitoring and Controlling, 9, Task 1

11. c. −$350

CV is calculated by EV − AC, or $1,500(2/3) − $1,350 = −$350. It is used to determine the amount of budget deficit (as in this example) or surplus at any given point in time. In this example, the project is over the planned cost; a positive value indicates it is under the planned costs, while a neutral value shows it on the planned costs. [Monitoring and Controlling]

PMI®, *PMBOK® Guide*, 2017, 262
PMI® *PMP Examination Content Outline*, 2015, Monitoring and Controlling, 9, Task 1

12. a. Cost management plan

The management and control of costs focuses on control thresholds or variance thresholds. Certain variances are acceptable, and others, usually those falling outside a particular range, are unacceptable. They are typically expressed as percentage deviations from the baseline plan. The actions taken by the project manager for variances are described in the cost management plan, which is the output of the Plan Cost Management Plan process. [Planning]

PMI®, *PMBOK® Guide*, 2017, 239
PMI® *PMP Examination Content Outline*, 2015, Planning, 6, Task 3

13. d. The information is insufficient to make an assessment.

   The information provided tells us that, as of the fourth month, more money has been spent than was planned. However, we need to know how much work has been completed to determine how the project is performing. In earned value terms, we are missing the EV or the measure of work expressed in terms of the budget authorized for the work. It is the budget associated for the work that has been completed. The objective is to establish progress measurement criteria for each WBS component to measure work in progress. EV then is monitored incrementally to determine current status and cumulatively to determine the long-term performance trends. [Monitoring and Controlling]

   PMI®, *PMBOK® Guide*, 2017, 261
   PMI® *PMP Examination Content Outline*, 2015, Monitoring and Controlling, 9, Task 1

14. b. Use the project schedule as an input to determine the budget

   Accurate project performance measurement depends on accurate cost and schedule information. The project schedule includes planned start and finish dates for all activities, milestones, and work packages. and control accounts. This information is used to aggregate costs to the calendar period for which the costs are planned to be incurred. The project schedule is an input under project documents in the Determine Budget process. [Planning]

   PMI®, *PMBOK® Guide*, 2017, 250
   PMI® *PMP Examination Content Outline*, 2015, Planning, 6, Task 3

15. d. Can be included in the activities or at higher levels

   Overhead or indirect costs include costs such as rent, insurance, or heating, which pertain to the project as a whole. These costs may be included in the activities or at a higher level if they are part of the project's cost estimate. The amount of overhead to be added to the project is frequently decided by the performing organization and may be beyond the control of the project manager. [Planning]

   PMI®, *PMBOK® Guide*, 2017, 246
   PMI® *PMP Examination Content Outline*, 2015, Planning, 6, Task 3

16. d. ETC = EAC − AC

    The ETC is the expected cost to finish the remaining work. This formula is used if the work is performing as planned. Otherwise you may need to re-estimate the remaining work from the bottom up using the formula of ETC = Re-estimate. [Monitoring and Controlling]

    PMI®, *PMBOK® Guide*, 2017, 267
    PMI® *PMP Examination Content Outline,* 2015, Monitoring and Controlling, 9, Task 1

17. c. Rough-order-of-magnitude

    A rough-order-of-magnitude estimate has an accuracy range of −25% to 75% and is made without detailed data. These estimates are when the project is in the initiation stage, but the cost estimates should be reviewed and refined during the project to reflect additional detail as the accuracy of the estimates will increase. Later in the project, for example definitive estimates can narrow the range of accuracy to −5% to +10%. [Planning]

    PMI®, *PMBOK® Guide*, 2017, 241
    PMI® *PMP Examination Content Outline,* 2015, Planning, 6, Task 3

18. b. Issue a change request

    Change requests are an output of the Control Costs process. Before a revised cost baseline leading to a budget update can be prepared, it is necessary to issue a change request. These change requests then are reviewed and processed through the Perform Integrated Change Control process. Change requests are an output of the Control Costs process and may be needed to other components in the project management plan. [Monitoring and Controlling]

    PMI®, *PMBOK® Guide*, 2017, 269
    PMI® *PMP Examination Content Outline,* 2015, Monitoring and Controlling, 9, Task 1

19. a. Add these opportunities to your reserves for later risk responses

    Reserve analysis is a tool and technique in the Control Costs process. Reserve analysis is used to monitor and control the status of contingency and management reserves for the project to see if they are still needed or if additional funds are required. However, the question notes opportunities, which tend to result in cost savings. When this is the case, funds may be added to the contingency reserve amount or taken from the project as a margin/profit. [Monitoring and Controlling]

    PMI®, *PMBOK® Guide*, 2017, 265
    PMI® *PMP Examination Content Outline,* 2015, Monitoring and Controlling, 9, Task 1

20. a. CPI

   The CPI has been proven to be an accurate and reliable forecasting tool. It is a measure of the cost efficiency of budgeted resources and is considered to be the most critical earned value measurement metric and measures the cost efficiency for the work completed. Its indices are useful for providing a basis to estimate project cost and schedule outcomes. [Monitoring and Controlling]

   PMI®, *PMBOK® Guide,* 2017, 263
   PMI® *PMP Examination Content Outline,* 2015, Monitoring and Controlling, 9, Task 1

21. b. Performance measurement baseline

   The performance measurement baseline is the approved plan for the project work, which is used to compare actual project execution. From it, deviations are measured for management control. It integrates scope, schedule, and cost parameters. It is an output of the Control Costs process as changes to it are incorporated in response to approved changes in scope, schedule performance, or cost control. If these variances are so severe, then a change request is used to revise the performance measurement baseline, so it is more realistic. [Monitoring and Controlling]

   PMI®, *PMBOK® Guide,* 2017, 269
   PMI® *PMP Examination Content Outline,* 2015, Monitoring and Controlling, 9, Task 1

22. b. Parametric estimating

   Parametric estimating involves using statistical relationships between historical data and other variables, such as square footage as in this example, to calculate a cost estimate for project work. This approach can produce higher levels of accuracy depending on the sophistication and underlying data in the model. Parametric estimates may be used for the entire project or for parts of it in conjunction with other estimating techniques. [Planning]

   PMI®, *PMBOK® Guide,* 2017, 244
   PMI® *PMP Examination Content Outline,* 2015, Planning, 6, Task 3

23. d. Costs that have been authorized and incurred

    These costs are part of work performance data about project progress. Work performance data are an input to the Control Costs process. In addition, these data include costs that have been invoiced and paid. If it is determined that the budget requires an update, knowledge about the actual costs spent to date is required, and any budget changes are approved according to the Perform Integrated Change Control process. [Monitoring and Controlling]

    PMI®, *PMBOK® Guide*, 2017, 260
    PMI® *PMP Examination Content Outline*, 2015, Monitoring and Controlling, 9, Task 1

24. a. Control account

    Rules of earned value performance measurement are part of the cost management plan and may (1) define the points in the WBS where measurement of control accounts will be performed; (2) establish the EV measurement techniques; and (3) state specific tracking methods and EV equations for calculating the EAC forecasts to provide a validity check on the bottom-up EAC. [Planning]

    PMI®, *PMBOK® Guide*, 2017, 265
    PMI® *PMP Examination Content Outline*, 2015, Planning, 6, Task 3

25. a. $EAC = ETC - AC$

    EAC is the expected total cost of completing all work expressed as the sum of the actual cost to date and the estimate at completion. There are four methods to compute it based on different assumptions. If the CPI is expected to be the same for the remainder of the project use $EAC = BAC/CPI$ or answer b. If the initial plan no longer is valid, use $EAC = AC + Bottom\text{-}up\ ETC$ or answer c. If both the CPI and the SPI influence the remaining work, use $EAC = AC + (BAC - EV)/(CPI \times SPI)$ or answer d. [Monitoring and Controlling]

    PMI®, *PMBOK® Guide*, 2017, 264–265
    PMI® *PMP Examination Content Outline*, 2015, Monitoring and Controlling, 9, Task 1

26. a. Schedule

   The project schedule is an input to the Estimate Costs process as part of project documents. The schedule is significant as it includes the type, quantity, and amount of time the team and physical resources will be used on the project. The duration estimates effect cost estimates when resources are charged one unit of time and when there are seasonal fluctuations in costs. It also provides useful information for projects that incorporate the cost of financing including any interest charged. [Planning]

   PMI®, *PMBOK® Guide*, 2017, 242
   PMI® *PMP Examination Content Outline*, 2015, Planning, 6, Task 3

27. b. PV, EV, and AC

   Cumulative cost curves, or S-curves, enable the project manager to monitor the three parameters of planned value, earned value, and actual cost and report on a period-by period basis such as weekly or monthly or on a cumulative basis. [Monitoring and Controlling]

   PMI®, *PMBOK® Guide*, 2017, 263
   PMI® *PMP Examination Content Outline*, 2015, Monitoring and Controlling, 8, Task 1

28. d. Represent the basic level at which project performance is measured and reported

   The points in the WBS where measurements of the control accounts will be performed are part of rules of performance measurement in the cost management plan. Control accounts also are used as the WBS component to link to the project's cost accounting system as stated in the section of this plan on organizational procedures links. They represent a management control point where scope, budget, actual costs, and schedule are integrated and compared to earned value for performance measurement. [Planning]

   PMI®, *PMBOK® Guide*, 2017, 239
   PMI® *PMP Examination Content Outline*, 2015, Planning, 6, Task 3

29. a. The Estimate at Completion

    Cost forecasts are an output of the Control Costs process. Either a calculated Estimate at Completion or a bottom-up EAC value is documented and communicated to stakeholders. The EAC is prepared as the project progresses to see if it is different from the budget at completion based on project performance. [Monitoring and Controlling]

    PMI®, *PMBOK® Guide*, 2017, 264, 269
    PMI® *PMP Examination Content Outline*, 2015, Monitoring and Controlling, 9, Task 1

30. a. Before the budget is complete

    Often project cost estimates are prepared after budgetary approval is provided. However, activity cost estimates should be prepared before the budget is complete. Cost estimates are a project document that are an input to the Determine Budget process. These cost estimates are for each activity within a work package that are aggregated for a cost estimate for the work package. [Planning]

    PMI®, *PMBOK® Guide*, 2017, 250
    PMI® *PMP Examination Content Outline*, 2015, Planning, 6, Task 3

31. c. Use decision making

    In Estimate Costs, one tool and technique is decision making. In it, voting is used as an assessment process with multiple alternatives with an expected outcome in the form of future actions. These techniques help engage team members to improve estimating accuracy and to promote commitment to the emerging estimates. [Planning]

    PMI®, *PMBOK® Guide*, 2017, 246
    PMI® *PMP Examination Content Outline*, 2015, Planning, 6, Task 3

32. d. Request additional contingency reserves for your budget

    Reserve analysis monitors the status of contingency and management reserves to see if they are still needed or if additional reserves need to be requested. The reserves may be used as planned to cover risk response costs or for other contingencies. In this situation risk identification and analysis are ongoing, and they may indicate a need to request additional reserves. [Monitoring and Controlling]

    PMI®, *PMBOK® Guide*, 2017, 265
    PMI® *PMP Examination Content Outline*, 2015, Monitoring and Controlling, 9, Task 1

33. b. Global supply and demand conditions can influence resource costs

    Market conditions, along with published commercial information, and exchange rates and inflation are enterprise environmental factors, an input to Estimate Costs. Market conditions describe the products, services, and results available in the market and who provides them along with any terms and conditions. As well, regional and global supply and demand conditions can greatly influence resource costs, assisting in the estimating process. [Planning]

    PMI® *PMBOK® Guide*, 2017, 243
    PMI® *PMP Examination Content Outline*, 2015, Planning, 6, Task 3

34. a. CVs are severe, and a realistic measure of performance is needed

    After the CVs exceed certain ranges, the original project budget may be questioned and changed as a result of new information. Changes to the cost baseline are incorporated in response to approved changes in scope, activity resources, or cost estimates. [Monitoring and Controlling]

    PMI®, *PMBOK® Guide*, 2017, 269
    PMI® *PMP Examination Content Outline*, 2015, Monitoring and Controlling, 8, Task 1

35. d. Document lessons learned

    Lessons learned but not documented are "lessons lost." The lessons learned register will help current project members, as well as people on future projects, make better decisions. Accordingly, the reasons for the variance, the rationale supporting the corrective action, and other related information must be documented. Further, the techniques that were effective in maintaining the budget, variance analysis, earned value analysis, forecasts, and corrective actions to respond to cost variance are documented. [Monitoring and Controlling]

    PMI®, *PMBOK® Guide*, 2017, 270
    PMI® *PMP Examination Content Outline*, 2015, Monitoring and Controlling, 9, Task 6

36. a. Aggregate these estimates by work packages

Cost estimates are a project document, which is an input to the Determine Budget process. As the budget is determined, the cost estimates for the activities should be aggregated by the work packages in the WBS. Ultimately, the cost baseline is developed, which is an approved version of the time-based budget excluding any management reserves, and it summarizes the approved budgets for the schedule activities. [Planning]

PMI®, *PMBOK® Guide*, 2017, 250, 254
PMI® *PMP Examination Content Outline*, 2015, Planning, 6, Task 3

37. a. All future work will need to be performed immediately with the range of the TCPI.

The TCPI is a measure of cost performance that is required to be achieved with the remaining resources to meet a specified management goal. It is expressed as the ratio of the cost to finish the remaining work to the remaining budget. If the cumulative CPI falls below the baseline, future project work will need to be performed immediately within the range of the TCPI to stay within the authorized budget at completion. Whether this performance level is achievable is a judgment call based on considerations such as risk, time remaining, and technical performance [Monitoring and Controlling]

PMI®, *PMBOK® Guide*, 2017, 226
PMI® *PMP Examination Content Outline*, 2015, Monitoring and Controlling, 9, Task 1

38. d. Consider the forecasted EAC

Forecasting is part of trend analysis, which is a tool and technique in Control Costs. In this situation, use of the forecasted EAC is recommended as it involves making projections of conditions and events in the future of the project based on current performance information and other knowledge that is available. Forecasts are generated, updated, and reissued based on work performance data provided as the project is executed. EACs are typically based on the actual costs for work completed plus an estimate of the remaining work. [Monitoring and Controlling]

PMI®, *PMBOK® Guide*, 2017, 264
PMI® *PMP Examination Content Outline*, 2015, Monitoring and Controlling, 9, Task 1

39. b. At –$200, the physical progress is being accomplished at a slower rate than is planned, indicating an unfavorable situation.

    Schedule variance is calculated: EV – PV or $1,000 – $1,200 = –$200. Because the SV is negative, physical progress is being accomplished at a slower rate than planned. It is a useful metric as it can indicate when a project is falling behind or is ahead of its baseline schedule. Ultimately, it equals zero when the project is complete because all the planned values have been earned. It should be used along with monitoring the critical path. [Monitoring and Controlling]

    PMI®, *PMBOK® Guide*, 2017, 262
    PMI® *PMP Examination Content Outline*, 2015, Monitoring and Controlling, 9, Task 1

40. d. Place emphasis on improving the productivity by which work was being performed

    CPI = EV/AC and measures the efficiency of the physical progress accomplished compared to the baseline. A CPI of 0.84 means that for every dollar spent, you are only receiving 84 cents of progress. Therefore, you should focus on improving the productivity by which work is being performed as now it represents a cost overrun for the project. The CPI is considered the most critical earned value metric. [Monitoring and Controlling]

    PMI®, *PMBOK® Guide*, 2017, 263
    PMI® *PMP Examination Content Outline*, 2015, Monitoring and Controlling, 9, Task 1

Project Quality Management

Practice Questions

INSTRUCTIONS: Note the most suitable answer for each multiple-choice question in the appropriate space on the answer sheet.

1. Quality is very important to your company. Each project has a quality statement that is consistent with the organization's vision and mission. In both internal and external Manage Quality, the broader definition of quality assurance, is provided on projects to—

   a. Ensure confidence that the project will satisfy relevant quality standards
   b. Monitor specific project results to note whether they comply with relevant quality standards
   c. Identify ways to eliminate causes of unsatisfactory results
   d. Use inspection to keep errors out of the process

2. Benchmarking is a technique used in—

   a. Inspections
   b. Root cause analysis
   c. Plan Quality Management
   d. Control Quality

3. In quality management, you plan to conduct tests for objective information about the product or service under test according to project requirements. Your intent in conducting these tests is to is—

   a. Ensure the product is acceptable under certain circumstances
   b. Examine the work products to see if they conform to requirements
   c. Take action to bring a defective or nonconforming component into compliance
   d. Find errors, defects, bugs, or other nonconformance problems

4. The requirements management plan is useful in Plan Quality Management because it—

   a. Provides better product definition and product development
   b. Helps document stakeholder needs and expectations
   c. Helps plan how quality control will be implemented on the project
   d. Helps manage requirements the quality management plan will reference

5. Assume you are working to Manage Quality on your project. You decided you wanted to select a tool that could help result in cost reduction and quality improvement so you used—

   a. Process decision point program charts
   b. A roadmap
   c. Design for X
   d. Force field analysis

6. You are leading a research project that will require between 10 and 20 aerospace engineers. Some senior-level aerospace engineers are available. They are more productive than junior-level engineers, who cost less and who are available as well. You want to determine the optimal combination of senior- and junior-level personnel. In this situation, the appropriate technique to use is to—

   a. Conduct a design of experiments
   b. Use the Ishikawa diagram to pinpoint the problem
   c. Prepare a control chart
   d. Analyze the process using a Pareto diagram

7. Check sheets are often called tally sheets. They are useful because they—

   a. Can be used to formulate the problem for a cause-and-effect diagram
   b. Use statistical techniques to compute a "loss function" to determine the cost of producing products that fail to achieve a target value
   c. Can be displayed in Pareto diagrams
   d. Can organize facts to help solve quality problems

8. One area that often influences perceptions of quality is—

   a. Cultural concerns
   b. Information from the risk register
   c. Organizational quality policies
   d. Specific tolerances

9. In Control Quality, it is useful to recognize when to use certain concepts, one of which is attribute sampling. Its concern is to—

a. Concentrate on prevention
b. Focus on conformance
c. Identify special causes
d. Determine tolerances

10. Your project scheduler has just started working with your project and has produced defective reports for the past two accounting cycles. If this continues, these defective reports could provide the potential for customer dissatisfaction and lost productivity that is due to rework. You discovered that the project scheduler needs additional training on using the scheduling tool that is used on your project. The cost of training falls under which one of the following categories?

a. Overhead costs
b. Failure costs
c. Prevention costs
d. Indirect costs

11. In order to use approved change requests as an input to Control Quality, it is important to—

a. Have a change management plan
b. Update the change log
c. Implement recommendations from the CCB
d. Verify their timely implementation

12. As you work to prepare your quality management plan for your project, you want to be able to determine which project deliverables and processes will require quality review. Therefore, you should—

a. Consider the organization's structure
b. Use statistical sampling
c. Review the scope baseline
d. Use check sheets

13. Assume you are working on preparing your project' quality management plan. As you prepare this plan, you and your team do not want to perform any unneeded rework, and you also want to keep the cost of quality low. Additionally, you want to meet the product's performance and reliability goals. Therefore, in preparing your plan, you should consider—

a. Trend analysis
b. Tests and inspections
c. Control charts
d. Mind maps

14. Project quality management was once thought to include only inspection or quality control. In recent years, the concept of project quality management has broadened. Failure to meet quality requirements can lead to—

    a. Employee attrition
    b. Quality Assurance having complete responsibility
    c. Customers requiring a documented and registered quality assurance system
    d. Unstable processes

15. Assume your organization is a start-up company, and you are trying to explain the importance of project quality management to the management team, which has not worked in this area before. You held an orientation session for them as to its importance. Then, you prepared your quality management plan and asked the management team to review it before it was completed and distributed to stakeholders. The purpose of their review was to—

    a. Have a sharper focus on the project's value proposition
    b. Ensure the team can provide needed quality metrics
    c. Determine the project's sufficient degree of accuracy and precision
    d. Commit to roles and responsibilities

16. Your quality assurance department recently performed a quality audit of your project and identified a number of findings and recommendations. One recommendation seems critical and should be implemented because it affects successful delivery of the product to your customer. Your next step should be to—

    a. Call a meeting of your project team to see who is responsible for the problem
    b. Reassign the team member who had responsibility for oversight of the problem
    c. Perform product rework immediately
    d. Issue a change request to implement accepted audit recommendations

17. One way to monitor cost and schedule variances, volume, frequency of scope changes, and other results to determine if the project management processes are in control is to use control charts. Assume you are working on a project that implements a repetitive process as the result is used many times after the process is designed and tested in telecommunications. Since it is repetitive, the control limits are

    a. Seven consecutive plot points above the mean
    b. Ways to identify which factors have the greatest influence
    c. +/− 3 s around a process mean set at 0 s
    d. Seven consecutive plot points below the mean

18. Your project is intended to result in a new manufacturing process in your company as the current process has remained the same for the past 20 years. You are preparing your project's quality management plan. You know your key stakeholders prefer to see the reports and other data visually so as you prepare this plan you are using—

   a. Force field analysis
   b. Logical data models
   c. Matrix diagrams
   d. The SIPOC model

19. Assume previous projects in the organization have overrun their budgets consistently and tend to require more contingency reserves than in the original budget. You are striving to avoid the need for additional contingency on your project and are doing so by—

   a. Involving more stakeholders in the risk identification process
   b. Using a process decision program chart
   c. Consulting the knowledge transfer system
   d. Using cost/benefit analysis

20. You want to prioritize the quality metrics as you are working in Plan Quality Management. An effective approach is to use—

   a. Problem solving
   b. Multicriteria decision analysis
   c. Interviews
   d. Checklists

21. Your management has prescribed that a quality audit be conducted at the end of every phase in a project. This audit is part of the organization's—

   a. Manage Quality process
   b. Control Quality process
   c. Quality Assurance program
   d. Process adjustment program

22. You are managing a major international project, and your contract requires you to prepare both a project plan and a quality management plan. Your core team is preparing a project quality management plan. You need to—

   a. Determine specific metrics to use in the quality management process
   b. Identify the quality standards for the project
   c. Use the organization's quality policies
   d. Identify specific quality management roles and responsibilities for the project

23. Recently your company introduced a new set of "metal woods" to its established line of golfing equipment. As you work in Plan Quality Management on this project, you decide to use a matrix diagram as it—

    a. Analyzes the product development cycle after product release to determine strengths and weaknesses
    b. Shows the strength of relationships between factors, causes, and objectives
    c. Identifies non-conformity, gaps, and shortcomings
    d. Analyzes the quality of the processes of the project against organizational standards

24. As you work to prepare your quality management plan and some quality metrics, you decide to create a single quality concept, which shows you are using—

    a. Mind mapping
    b. Flowcharts
    c. Design for X
    d. Problem solving

25. On-time performance, CPI, defect frequency, failure rates, defects identified each day, down time each month, and customer satisfaction scores are examples of—

    a. Incentives to vendors to make quality commitments to improve overall performance
    b. Quality metrics as an input to Manage Quality
    c. Methods that usually result in lower costs and increased profitability
    d. Items to include as goals in the quality management plan

26. Rework required, causes for rejection, or the need for process adjustment are examples of—

    a. Work performance data in Manage Quality
    b. Work performance information in Control Quality
    c. Items that are part of the process improvement plan
    d. Items to cover at meetings used to prepare the quality management plan

27. Quality control measurements are captured—

    a. As stated in the quality management plan
    b. To prepare an operational definition
    c. To prepare a control chart
    d. As an output of Manage Quality

28. Assume you are working to Manage Quality on your project. You are focusing on improving the quality management methods you are using. One approach to consider is—

    a. Monitoring process variation over time
    b. Using Six Sigma
    c. Determining whether results conform to benefits
    d. Using Design for X

29. An often, but important, overlooked output of the Manage Quality is used by other process and departments to take corrective action. This output is—

    a. The process improvement plan
    b. Benefits delivery management
    c. Quality reports
    d. Work performance information

30. You are a project manager for residential construction. As a project manager, you must be especially concerned with building codes—particularly in the Plan Quality Management process. You must ensure that building codes are reflected in your project plans because—

    a. Standards and regulations are an input to Plan Quality Management
    b. Quality audits serve to ensure there is compliance with regulations
    c. They are a cost associated with quality initiatives
    d. Compliance with standards is the primary objective of Control Quality

31. You work as a project manager in the largest hospital in the region. Studies have shown that patients have to wait for long periods before being treated. To assist in identifying the factors contributing to this problem, you and your team have decided to use which of the following techniques?

    a. Cause-and-effect diagrams
    b. Pareto analysis
    c. Scatter diagrams
    d. Control charts

32. Assume you are working to prepare your quality management plan as it is critical to success in your construction company. You realize as you do so that it would be helpful to have an overview of the tests required to verify requirements, so you decide to—

    a. Use a checklist
    b. Review the requirements traceability matrix
    c. Have a framework for needed quality systems
    d. Review assumptions and constraints

33. You have prepared your quality management plan and identified quality metrics. Now you are working to Manage Quality on your project. You want to know why you have a variance and some defects, so you decide to—

   a. Use an affinity diagram
   b. Identify gaps and shortcomings
   c. Use cause-and-effect analysis
   d. Use root cause analysis

34. A tool used to verify that a set of required steps have been performed and incorporates acceptance criteria is—

   a. A check sheet
   b. Problem solving
   c. A checklist
   d. Process analysis

35. One way to display the sequence of steps and the branching possibilities that exist for a process that transforms one or more inputs into one or more outputs is to use a process map or a—

   a. Checklist
   b. Flowchart
   c. Tree diagram
   d. Process decision program chart

36. Quality objectives of the project are recorded in—

   a. Process improvement plan
   b. Quality management plan
   c. Quality baseline
   d. Quality metrics

37. The Manage Quality process consists of planned and systematic acts and processes. They help to—

   a. Determine the cost of quality.
   b. Implement specific design guidelines
   c. Prepare quality metrics
   d. Conduct inspections

38. You have decided to use a fishbone diagram to identify the relationship between an effect and its causes. To begin, you should first—

    a. Select an interdisciplinary team who has used the technique before to help brainstorm the problem
    b. Determine the major categories of defects
    c. Set up a process analysis using charts
    d. Identify the problem

39. Assume you want to show the relationship between two variables. You then want to use a(n)—

    a. Histogram
    b. Attribute chart
    c. Control chart
    d. Scatter diagram

40. Processes often interact with one another. For example, the output of a process may be an input to another process in a different knowledge area. An example from quality management is—

    a. Verified deliverables
    b. Realized benefits
    c. Business value
    d. Process metrics

Answer Sheet

	a	b	c	d			a	b	c	d
1.	a	b	c	d		21.	a	b	c	d
2.	a	b	c	d		22.	a	b	c	d
3.	a	b	c	d		23.	a	b	c	d
4.	a	b	c	d		24.	a	b	c	d
5.	a	b	c	d		25.	a	b	c	d
6.	a	b	c	d		26.	a	b	c	d
7.	a	b	c	d		27.	a	b	c	d
8.	a	b	c	d		28.	a	b	c	d
9.	a	b	c	d		29.	a	b	c	d
10.	a	b	c	d		30.	a	b	c	d
11.	a	b	c	d		31.	a	b	c	d
12.	a	b	c	d		32.	a	b	c	d
13.	a	b	c	d		33.	a	b	c	d
14.	a	b	c	d		34.	a	b	c	d
15.	a	b	c	d		35.	a	b	c	d
16.	a	b	c	d		36.	a	b	c	d
17.	a	b	c	d		37.	a	b	c	d
18.	a	b	c	d		38.	a	b	c	d
19.	a	b	c	d		39.	a	b	c	d
20.	a	b	c	d		40.	a	b	c	d

Answer Key

1. a. Ensure confidence that the project will satisfy relevant quality standards

    Manage Quality, the broader term than quality assurance, increases project effectiveness and efficiency or project activities so better results are achieved and performance increases to enhance stakeholder satisfaction. It implements the planned and systematic acts and processes in the project's quality management plan. It builds confidence that future output progress will be completed in a way to meet specified requirements and specifications. Manage Quality should be performed throughout the project. [Executing]

    PMI®, *PMBOK® Guide*, 2017, 288–289
    PMI® *PMP Examination Content Outline*, 2015, Executing, 8, Task 3

2. c. Plan Quality Management

    Benchmarking involves comparing actual or planned practices to those practices or quality standards of comparable projects to identify best practices, to note ideas for improvement, and to provide a way to measure performance. Benchmarking may be done by the performing organization or external to it or can be within the same application area. It enables analogies from projects in different application areas to be made. It is a tool and technique in the data gathering category of Plan Quality Management. [Planning]

    PMI®, *PMBOK® Guide*, 2017, 281
    PMI® *PMP Examination Content Outline*, 2015, Planning, 5, Task 8

3. d. Find errors, defects, bugs, or other nonconformance problems

    Testing is a tool and technique used in Control Quality. Early testing helps identify nonconformance problems and reduces the cost of fixing them. The intent of testing is the answer to this question. The type, amount, and extent of testing is in the quality management plan and depends on the nature of the project. Tests can be performed throughout the project if desired. [Monitoring and Controlling]
    PMI®, *PMBOK® Guide*, 2017, 303

    PMI® *PMP Examination Content Outline*, 2015, Monitoring and Controlling, 9, Task 1

4. d. Helps manage requirements the quality management plan will reference

The requirements management plan is a project document used as an input to Plan Quality Management. It provides the approach to identify, analyze, and manage the requirements the quality management plan and the quality metrics will reference. [Planning]

PMI®, *PMBOK® Guide*, 2017, 279
PMI® *PMP Examination Content Outline*, 2015, Planning, 6, Task 8

5. c. Design for X

Design for X is a set of technical guidelines to use during product design to optimize a specific aspect of the design. It can control and hopefully improve the product's final characteristics. The 'X' means different aspects of product development that include reliability, deployment, assembly, manufacturing, cost, service, capability, safety, and quality. [Executing]

PMI®, *PMBOK® Guide*, 2017, 295
PMI® *PMP Examination Content Outline*, 2015, Executing, 8, Task 3

6. a. Conduct a design of experiments

Manage Quality is the work of everyone on the project. While the project manager and team may use the Quality Assurance Department to assist in this process, activities such as failure analysis, design of experiments, and quality improvement are stressed. This situation in this question is an example of designing an experiment. By analyzing the data, it can help provide the optimal conditions for the product or process, highlight the factors that influence the results, and reveal the presence of interactions and synergy among the factors. [Executing]

PMI®, *PMBOK® Guide*, 2017, 290
PMI® *PMP Examination Content Outline*, 2015, Executing, 8, Task 3

7. d. Can organize facts to help solve quality problems

Check sheets are a tool and technique in Control Quality. They are useful to organize facts to facilitate collecting relevant data about a quality problem. Their use increases in value for gathering attribute data while performing inspections to identify defects. An example is the frequencies or consequences of defects collected. [Monitoring and Controlling]

PMI®, *PMBOK® Guide*, 2017, 302
PMI® *PMP Examination Content Outline*, 2015, Monitoring and Controlling, 9, Task 3

8. a. Cultural concerns

    Enterprise environmental factors are an input to Plan Quality
    Management, one of which is cultural perceptions that may influence
    expectations about quality. Other enterprise environmental factors
    are: government agency regulations; rules, standards, and guidelines
    specific to the application area of the project; geography, organizational
    structure, marketplace conditions; and working or operating conditions
    of the project or its deliverables that may affect product quality.
    [Planning]

    PMI®, *PMBOK® Guide*, 2017, 280
    PMI® *PMP Examination Content Outline*, 2015, Planning, 6, Task 8

9. b. Focus on conformance

    Attribute sampling determines whether a result does or does not
    conform. The project team ideally should have a working knowledge of
    some aspects of statistical sampling, one of which is attribute sampling.
    They also should know about the difference between prevention
    and inspection, variable sampling, and tolerances. [Monitoring and
    Controlling]

    PMI®, *PMBOK® Guide*, 2017, 274
    PMI® *PMP Examination Content Outline*, 2015, Monitoring and
    Controlling, 9, Task 3

10. c. Prevention costs

    Prevention costs include any expenditures directed toward ensuring
    that quality is achieved the first time. They include costs related to poor
    quality in products, deliverables, services, or the entire project. The
    objective is to focus on cost of conformance and spend money during
    the project to avoid failures. They are part of the cost of quality, a tool
    and technique in Plan Quality Management. The cost of conformance
    includes these prevention costs and appraisal costs. [Planning]

    PMI®, *PMBOK® Guide*, 2017, 282–283
    PMI® *PMP Examination Content Outline*, 2015, Planning, 6, Task 8

11. d. Verify their timely implementation

Approved change requests are an input to Control Quality. As part of the Perform Integrated Change Control process, a change log update shows some changes are approved, while others are rejected or deferred. The approved change requests may include modifications such as defect repairs, revised work methods, and a revised schedule. Their timely implementation should be verified, confirmed for completeness, retested, and certified as correct. [Monitoring and Controlling]

PMI®, *PMBOK® Guide*, 2017, 301
PMI® *PMP Examination Content Outline*, 2015, Monitoring and Controlling, 9, Task 3

12. c. Review the scope baseline

The scope baseline is part of the project management plan and is an input to Plan Quality Management. It contains the WBS and the deliverables documented in the scope statement to determine which quality standards and objectives are needed for the project and the project deliverables and processes that will require quality review. The scope statement also includes the acceptance criteria for the project's deliverables. This definition of the acceptance criteria may influence quality costs, but by satisfying the acceptance criteria shows stakeholder needs are met. [Planning]

PMI®, *PMBOK® Guide*, 2017, 279
PMI® *PMP Examination Content Outline*, 2015, Planning, 6, Task 8

13. b. Tests and inspections

Test and inspection planning is a tool and technique in Plan Quality Management. By using it, the project manager and the team determine how to test or insect the product, deliverable, or service to meet stakeholder expectations and the goals of product performance and reliability. Tests and inspections can include alpha and beta tests in a software project, strength tests in a construction project, inspection in manufacturing, and field tests and nondestructive tests in engineering. [Planning]

PMI®, *PMBOK® Guide*, 2017, 285
PMI® *PMP Examination Content Outline*, 2015, Planning, 6, Task 8

14. a. Employee attrition

Failure to meet quality requirements can result in serious negative consequences. One example is employee attrition as many people may not be needed if quality requirements are not met. Also in this category are overworking the project team, leading to decreased profits and increased levels of project risks, errors, or rework. [Monitoring and Controlling]

PMI®, *PMBOK® Guide*, 2017, 273
PMI® *PMP Examination Content Outline*, 2015, Monitoring and Controlling, 9, Task 3

15. a. Have a sharper focus on the project's value proposition

The content, style, and level of detail of a quality management plan depend on the project. It should be reviewed early in the project to ensue later decisions are made with accurate information. By having a review, its benefits are a sharper focus on the project's value proposition, of interest to executive management, cost reductions, and fewer schedule overruns as a result of rework. [Planning]

PMI®, *PMBOK® Guide*, 2017, 286
PMI® *PMP Examination Content Outline*, 2015, Planning, 6, Task 8

16. d. Issue a change request to implement accepted audit recommendations

The information obtained from a quality audit can be used to improve quality systems and performance. The subsequent effort to correct any deficiencies should result in a reduced cost of quality and an increase in sponsor or customer satisfaction. In most cases, quality audits also can confirm implementing approved change requests such as updates, corrective or preventive actions, and defect repairs. Change requests to implement recommendations from the quality audits are an output of the Manage Quality process. [Executing]

PMI®, *PMBOK® Guide*, 2017, 294–295
PMI® *PMP Examination Content Outline*, 2015, Executing, 8, Task 3

17. c. +/− 3 s around a process mean set at 0 s

    Control charts are a data representation tool and technique used in Control Quality The key words in this question are "repetitive process", which means the control limits are set at +/− 3 s around a process mean set at 0 s. The control charts are used to determine whether or not a process is stable or has predictable performance. Upper and lower limits are set based on requirements, and they reflect the maximum or minimum values allowed. These control limits are not specification limits. They are determined statistically. and the project manager and others can use the statistically calculated control limits to identify the points where corrective action is required. [Monitoring and Controlling]

    PMI®, *PMBOK® Guide*, 2017, 304
    PMI® *PMP Examination Content Outline,* 2015, Monitoring and Controlling, 9, Task 3

18. b. Logical data models

    In Plan Quality Management, logical data models are an example of a data representation tool and technique. A logical data model consists of a visual representation of an organization's data. Such a model is described in business language, so it can be used independently of specific technology. The logical data model also is useful as it can show where data integrity or other quality issues may arise. The key word in this question is visually. [Planning]

    PMI®, *PMBOK® Guide*, 2017, 284
    PMI® *PMP Examination Content Outline,* 2015, Planning, 6, Task 8

19. d. Using cost/benefit analysis

    Cost/benefit analysis is a data analysis tool and technique in Plan Quality Management. It is financial analysis tool, but here In Plan Quality Management, its purpose is to estimate strength and weaknesses of various alternatives in terms of benefits provided. This analysis helps the project manager evaluate if the planned quality activities will be effective. The benefits if quality requirements are met include less rework, lower costs, higher productivity, increased customer satisfaction and increased profits. This analysis for each quality activity compares the cost of quality to the expected benefits. [Planning]

    PMI®, *PMBOK® Guide*, 2017, 282
    PMI® *PMP Examination Content Outline,* 2015, Planning, 6, Task 8

20. b. Multicriteria decision analysis

The key words in this question are quality metrics. In the Plan Quality Management process, decision making is a tool and technique. Multi-criteria decision analysis tools, such as a prioritization matrix, are used to identify key issues and suitable alternatives to be prioritized as a set of decisions to implement. The criteria are analyzed and may be weighted to obtain a mathematical score for each alternative. Then, alternatives are ranked by their specific score. In this process, it helps prioritize quality metrics. [Planning]

PMI®, *PMBOK® Guide*, 2017, 283
PMI® *PMP Examination Content Outline*, 2015, Planning, 6, Task 8

21. a. Manage Quality process

Audits are a tool and technique in the Manage Quality process. Manage Quality translates the quality management plan into activities that incorporate the organization's quality polices into the project. Audits then are used to see if the project activities comply with organizational and project quality policies, processes, and procedures. They may be done by people external to the organization or by internal staff members. [Executing]

PMI®, *PMBOK® Guide*, 2017, 294
PMI® *PMP Examination Content Outline*, 2015, Executing, 8, Task 3

22. c. Use the organization's quality policies

In preparing your project's quality management plan, organizational processes assets should be considered. One of them, which is the key to this question, is the organizational quality management system, which includes policies, procedures, and guidance, among other things. The quality policy is the overall intentions and direction of the organization with regard to quality, as formally expressed by top management. [Planning]

PMI®, *PMBOK® Guide*, 2017, 281
PMI® *PMP Examination Content Outline*, 2015, Planning, 6, Task 8

23. b. Shows the strength of relationships between factors, causes, and objectives

    Matrix diagrams are an example of a data representation tool and technique used in Plan Quality Management. They show the strengths of relationships among different factors, causes, and objectives that are displayed in the rows and columns in the matrix. The project manager then determines how many factors to be compared to determine the specific shape of the matrix diagram. The diagram helps identify the key quality metrics important for project success. Quality metrics are an output of this process. [Planning]

    PMI®, *PMBOK® Guide*, 2017, 284
    PMI® *PMP Examination Content Outline*, 2015, Planning, 6, Task 8

24. a. Mind mapping

    Mind mapping is a data representation tool and technique used in Plan Quality Management. It is a diagrammatic method in which one can visualize organizing information. In quality, a mind map is created around a single quality concept, which is drawn as an image in the center of a blank page - usually in landscape. Then, associated represented ideas such as images, words, and parts of words are added. This technique often results in a quick way to gather quality requirements, constraints, dependencies, and relationships. [Planning]

    PMI®, *PMBOK® Guide*, 2017, 284
    PMI® *PMP Examination Content Outline*, 2015, Planning, 6, Task 8

25. b. Quality metrics as an input to Manage Quality

    These examples of metrics describe a project or product attribute and how the Control Quality process will verify compliance. In the Manage Quality process, these metrics are used as a basis for test scenarios for the project and its deliverables and as a basis for areas of improvement. [Executing]

    PMI®, *PMBOK® Guide*, 2017, 291
    PMI® *PMP Examination Content Outline*, 2015, Executing, 8, Task 3

26. b. Work performance information in Control Quality

In the Control Quality process, work performance information is an output. It is the information about project requirements fulfillment as shown in the answer to the question. Other examples are causes for rejection, needed rework, corrective action recommendations, lists of verified deliverables, status of quality metrics, and the need for process adjustments. [Monitoring and Controlling]

PMI®, *PMBOK® Guide*, 2017, 305
PMI® *PMP Examination Content Outline*, 2015, Monitoring and Controlling, 9, Task 3

27. a. As stated in the quality management plan

One Control Quality output results in quality control measurements. They are the documented results of the activities in Control Quality. They should be captured in the format specified in the quality management plan since this plan states planned quality control and quality management activities for the project, among other things [Monitoring and Controlling]

PMI®, *PMBOK® Guide*, 2017, 286, 305
PMI® *PMP Examination Content Outline*, 2015, Monitoring and Controlling, 9, Task 3

28. b. Using Six Sigma

In the Manage Quality process, one tool and technique is quality measurement methods. Quality improvements often occur based on recommendations from quality control, findings from quality audits, or problem solving. Plan-do-check-act and Six Sigma are two common quality improvement tools used to analyze and evaluate improvement opportunities. [Executing]

PMI®, *PMBOK® Guide*, 2017, 296
PMI® *PMP Examination Content Outline*, 2015, Executing, 7, Task 3

29. c. Quality reports

Quality reports can be graphical, narrative, or quantitative. They serve to provide the information to be used by other processes and departments to take corrective actions to achieve quality expectations. Information in these reports may include any quality management issues escalated by the project team; recommendations for improvement in the project, processes, and product requirements; recommendations for corrective action, and summary findings from Control Quality. [Executing]

PMI®, *PMBOK® Guide*, 2017, 296
PMI® *PMP Examination Content Outline*, 2015, Executing, 7, Task 3

30. a. Standards and regulations are an input to Plan Quality Management

    During the Plan Quality Management process, the project management team should consider enterprise environmental factors especially when relevant to a specific application area such as standards and regulations in this question with the needed building codes. Other examples are rules and guidelines, geographic distribution, the organizational structure, marketplace conditions, working or operating conditions, and cultural perceptions that may influence quality expectations. These are conditions outside of the control of the project manager and his or her team. [Planning]

    PMI®, *PMBOK® Guide*, 2017, 280
    PMI® *PMP Examination Content Outline*, 2015, Planning, 6, Task 8

31. a. Cause-and-effect diagrams

    Cause-and-effect diagrams, also called Ishikawa diagrams or fishbone diagrams, are used to illustrate how various causes and sub causes interact to create a special effect. It is named for its developer, Kaoru Ishikawa. These diagrams break down the causes of a problem statement into distinct branches. The objective is to identify the main or root cause of the problem. They are a data representation tool and technique in Manage Quality. [Executing]

    PMI®, *PMBOK® Guide*, 2017, 293–294
    PMI® *PMP Examination Content Outline*, 2015, Executing, 8, Task 3

32. b. Review the requirements traceability matrix

    The requirements traceability matrix is a project document that is an input to Plan Quality Management. It is helpful because it shows the links from product requirements to deliverables, and it helps ensure each requirement is tested. It also provides an overview of the tests needed to verify the requirements. [Planning]

    PMI®, *PMBOK® Guide*, 2017, 280
    PMI® *PMP Examination Content Outline*, 2015, Planning, 6, Task 8

33. d. Use root cause analysis

    Root cause analysis is a data analysis tool and technique in Manage Quality. Its purpose is to serve as an analytical approach to determine the basic underlying reason that causes a defect, variance, or risk. A single root cause may underlie more than one variance, defect, or risk. It also can be used to identify the root causes of a problem and then solving them. Once all root causes of a problem are resolved, the problem then does not exist. [Executing]

    PMI®, *PMBOK® Guide*, 2017, 292
    PMI® *PMP Examination Content Outline*, 2015, Executing, 8, Task 3

34. c. A checklist

Checklists are a data gathering tool and technique used in Manage Quality. They are a structured tool used to verify a set of steps has been performed or to see if requirements are satisfied. They may be basic or complex based on the project requirements. Quality checklists also incorporate the acceptance criteria in the scope baseline. [Executing]

PMI®, *PMBOK® Guide*, 2017, 292
PMI® *PMP Examination Content Outline*, 2015, Executing, 8, Task 3

35. b. Flowchart

A flowchart is a data representation technique used in Plan Quality Management and also in Manage Quality. In Plan Quality Management, they are also called process maps as noted in the question. They show the activities, decision points, branching loops, parallel paths, and the overall order of processing by mapping the organizational details of procedures that exist within a horizontal chain known as SIPOC or supplier, input, process, outputs, and customers model. [Planning]

PMI®, *PMBOK® Guide*, 2017, 284
PMI® *PMP Examination Content Outline*, 2015, Planning, 6, Task 8

36. b. Quality management plan

The quality management plan describes how policies, procedures, and guidelines will be implemented to achieve the quality objectives. It also includes quality standards, roles and responsibilities, deliverables and requirements that need review, planned quality control and quality management activities, quality tests, and procedures relevant to the project. [Planning]

PMI®, *PMBOK® Guide*, 2017, 286
PMI® *PMP Examination Content Outline*, 2015, Planning, 6, Task 8

37. b. Implement specific design guidelines

The Manage Quality process is important for many reasons. Using planned systematic acts and processes, it designs a mature and optimal product. It does so by implementing specific design guidelines. They address the specific aspects of the product. [Executing]

PMI®, *PMBOK® Guide*, 2017, 290
PMI® *PMP Examination Content Outline*, 2015, Planning, 6, Task 8

38. d. Identify the problem

The first and most important is to identify the problem as a gap to be closed or as an objective to be achieved. Causes then are found by looking at the problem statement and asking why until a root cause has been identified for which action can be taken or the reasonable possibilities on the diagram have been exhausted. [Executing]

PMI®, *PMBOK® Guide*, 2017, 295
PMI® *PMP Examination Content Outline*, 2015, Executing, 8, Task 3

39. d. Scatter diagram

Scatter diagrams are a data representation tool and technique in Manage Quality. The scatter diagram is a graph used to show the relationship between two variables. For example, they can show a relationship between any element of a process, environment, or activity on one axis and a quality defect on the other axis. [Executing]

PMI®, *PMBOK® Guide*, 2017, 293
PMI® *PMP Examination Content Outline*, 2015, Executing, 8, Task 3

40. a. Verified deliverables

An output of Control Quality is verified deliverables. A goal of this process is to determine the correctness of deliverables. Verified deliverables then are an input to the Validate Scope process for formalized acceptance. However, it is important to note that any change requests or improvements related to the deliverables means they then may be changed, inspected, and reviewed. [Monitoring and Controlling]

PMI®, *PMBOK® Guide*, 2017, 305
PMI® *PMP Examination Content Outline*, 2015, Monitoring and Controlling, 9, Task 3

Project Resource Management

Practice Questions

INSTRUCTIONS: Note the most suitable answer for each multiple-choice question in the appropriate space on the answer sheet.

1. You have been assigned as project manager on what could be a "bet the company" project. You realize that to be successful you need to exercise maximum control over project resources. You are both a manager and a leader in resource management. This means you must—

    a. Focus on emotional intelligence
    b. Model professional and ethical behavior
    c. Select the team resources with the needed competencies
    d. Recognize and promote diversity

2. As the project manager, it is your goal to build a high-performing team. On your project, four of your nine team members know one another, but five are new to you and the rest of the team. As you work in team building, your objective is to—

    a. Provide recognition and rewards
    b. Reach consensus on team issues
    c. Help team members work together effectively
    d. Try to solve team political problems

3. Assume you are managing a large construction project to use solar power for electricity in a different country in a location that lacks easy access to transportation facilities. You must make sure all the equipment and other needed materials are available when you need them. Given the geographical location, it is a difficult task. Much of the construction equipment also is leased. As a result you need to—

   a. Use a lean approach
   b. Use variance analysis
   c. Use agreements
   d. Emphasize marketplace conditions

4. You are leading a team to reorganize your IT department, so it is more customer oriented. Your team has 10 people. While your team members have competencies in IT, they are not in the department to promote objectivity. In your resource plan, you have a section on recognition to state when recognition and rewards will be given to members of your team. As you put this plan into practice, you should consider—

   a. Compatibility between the rewards you give and those in the team members' functional group
   b. Whether the team charter is followed
   c. Team motivation
   d. Cultural differences

5. Which of the following factors contributes the most to team communication?

   a. External feedback
   b. Performance appraisals
   c. Smoothing over team conflicts by the project manager
   d. Colocation

6. You are managing a virtual team. The project has been under way for several months, and you believe your team members do not view themselves as a team or a unified group. To help rectify this situation, you should—

   a. Ensure that every member of the project team uses the same type of e-mail system
   b. Enhance the use of technology
   c. Enhance communications planning
   d. Establish discussion forums

7. Major difficulties arise when multiple projects need to be managed in the functional organizational structure because of—

   a. The level of authority of the project manager
   b. Conflicts over the relative priorities of different projects in competition for limited resources
   c. Project team members who are focused on their functional specialty rather than on the project
   d. The need for the project manager to use interpersonal skills to resolve conflicts informally

8. Leadership is embedded in the job of the project manager and really in the job of every team member. Assume you are project manager with a team of two people who will work full-time on the project, and six people who will support the project on a part-time basis. All team members know one another and have worked together in the past. Leadership on this team is especially important—

   a. In the planning phase
   b. At the beginning of the project
   c. During the execution phase
   d. When phase gate and performance reviews are held

9. Your organization is characterized by hierarchical organizational structures with rigid rules and policies and strict supervisory controls. Individual team members are not expected to engage in problem solving or use creative approaches to plan and execute work; management has these responsibilities. Your organization is characterized by which one of the following theories?

   a. Ouchi's Theory
   b. McGregor's Theory X
   c. Maslow's self-esteem level
   d. Vroom's Expectancy Theory

10. As you prepare your resource management plan, you need to provide guidance and determine the skills and capacity required to complete the activities in the project. This should be documented in the—

    a. Roles and responsibilities section
    b. Staffing management plan
    c. Staff acquisition section
    d. Competencies section

11. The primary purpose of effective team development is—

    a. Enhanced project performance
    b. Motivated employees with reduced attrition rates
    c. Enhanced interpersonal skills
    d. Improved team work

12. The team members on your project have been complaining that they do not have any sense of identity as a team because they are located in different areas of the building. To remedy this situation, you developed a project logo and had it printed on T-shirts to promote the project, but this action has not worked. Your next step is to—

    a. Initiate a newsletter
    b. Establish a project vision
    c. Establish a team charter
    d. Encourage problem solving

13. You are managing a team of 120 internal members and four contractors. Given the size of this team, you want to document each team member's assignments. You can use a—

    a. Resource breakdown structure
    b. RACI chart
    c. Team calendar
    d. Team directory

14. Assume you are preparing your resource management plan. Your goal is to ensure each work package has an owner, who is clearly identified, and all team members understand their roles and responsibilities. The best way to show these data is by using a(n)—

    a. RAM
    b. Chart
    c. Resource breakdown structure
    d. Organizational breakdown structure

15. You are managing a design and construction project for the next-generation of drones. Your organization has been the leader in the drone business since it began, which means you are fortunate as your executive realize they would only increase in capabilities and popularity and acquired needed resources in house rather than rely on contractors to preserve intellectual property. On your project, you find you are in a dilemma as you need a specific type of crane from another department. If you do not get the crane when you need it, your project's schedule will slip dramatically. The best approach you can use in this situation is—

    a. Problem solving
    b. Political skills
    c. Referent power
    d. Lease the crane

16. You have been a project manager for seven years. You now are managing the construction of a new facility that must comply with the government's newly issued environmental standards. You want to ensure that your team members are able to select methods to complete various activities on the project without needing to involve you in each situation. As you prepare your resource management plan, you should document this information in which of the following—

    a. Roles and responsibilities section
    b. Authority section
    c. Resource breakdown structure
    d. Staffing management plan

17. It is important on all projects to determine when and how human resources will be met. Assume that you are managing a project to assess methods for streamlining the regulatory approval process for new medical devices in your government agency. Because the agency has undergone downsizing during the past three years, subject matter experts are in short supply. You must determine whether the needed subject matter experts can be acquired from inside the agency or whether you must use contractors. This information should be documented in the—

    a. Make-or-buy decisions in the procurement management plan
    b. Contracts management plan
    c. Staffing management plan
    d. Resource management plan

18. Conflicts are inevitable on projects, but if actually managed effectively, at times conflicts can help the team arrive at a better solution. One of the challenges a project manager faces is—

    a. Determining sources of conflict early and having proactive solutions to manage them
    b. Recognizing stakeholders will have conflicting interests but use a collaborative approach to resolve each conflict
    c. Resolve it in a timely way
    d. Managing conflict when it occurs

19. As project manager, you are primarily responsible for implementing the project management plan by authorizing the execution of project activities. Because you do not work in a projectized organization, you do not have direct access to human resource administrative activities. Therefore, you need to—

    a. Outsource these functions
    b. Include a human resource representative in your RAM chart
    c. Ensure that your team is sufficiently aware of administrative requirements to ensure compliance
    d. Have a representative from human resources sign off on the team charter

20. Constant bickering, absenteeism, and substandard performance have characterized the behavior of certain members of your team. You have planned an off-site retreat for the team to engage in a variety of activities. Your primary objective for investing time and money in this event is because your team—

    a. Is in the storming stage
    b. Has too many issues to solve
    c. Both individual and team performance are lagging
    d. Is in the forming stage

21. Two team members on your project often disagree. You need a conflict resolution method that provides a long-term resolution. You decide to use which one of the following approaches?

    a. Confronting
    b. Accommodating
    c. Collaborating
    d. Smoothing

22. Assume you are working to prepare your resource management plan for your automotive company. You are going to design and then manufacture state-of-the-art vehicles of various types to surpass your competitors who are focused on self-autonomous vehicles. As you prepare your plan, one area to consider is—

    a. The organizational structure of your company
    b. Your scope statement
    c. Talent management
    d. Reporting requirements

23. As a project manager, you believe in using a "personal touch" to further team development. One approach that has proven effective toward this goal is—

    a. Creating a team name
    b. Providing flexible work time
    c. Improving feelings of trust and agreement
    d. Using conversations with team members

24. Your project has been under way for some time with only a few direct reports. However, it then is your responsibility to—

    a. Facilitate an atmosphere conducive to success
    b. Focus on ensuring team member strengths are known and assign them to roles to enable the project to be completed as soon as possible
    c. Develop a recovery plan as a contingency to help complete the project
    d. Provide individual and team assessments

25. You are the project manager for a two-year project that is now beginning its second year. The mix of team members has changed, and there is confusion as to roles and responsibilities. In addition, several of the completed work packages have not received the required sign-offs, and three work packages are five weeks behind schedule. To gain control of this project, you need to—

    a. Rebaseline your original resource plan with current resource requirements
    b. Change to a projectized organizational structure for maximum control over resource assignments
    c. Work with your team to prepare a responsibility assignment matrix
    d. Create a new division of labor by assigning technical leads to the most critical activities

26. You are the project manager to develop a new medical implant device. You were selected to manage this project since you are an expert in medical implant devices. However, you strive to seek opinions from others. Now you are estimating the resources you will need for this project. You consulted with an estimating expert in the company and decided the best approach is to use—

    a. Analogous estimating
    b. Parametric estimating
    c. Expert judgment
    d. Meetings

27. Having estimated the resources you need for your new medical implant project, the next challenge is to acquire them. Resources seem to always be scarce in the company. Additionally, you also work in a matrix environment, so you have limited direct control over resource selection. Your best approach is to—

    a. Build on the meetings you held with the key people and use communications skills
    b. Build on the meetings you held with the key people and use influencing skills
    c. Review resource calendars and then use resource leveling
    d. Meet with peers to seek their opinions as to what they would do

28. Assume although you did your best in negotiating and influencing others to release resources to you for your medical implant project, you were not successful. You now are concerned your project, even though it had a strong business case and was approved by the portfolio group, that you now have another constraint to consider, which is—

    a. Pre-assignment of staff to the project
    b. Economic factors
    c. Use of outsourcing
    d. Team member training requirements

29. Decision making often is used as a tool and technique in Acquire Project Team. One approach is multi-criteria decision tool. An example of a selection criterion to consider is—

    a. Competency
    b. Motivation
    c. Attitude
    d. Trust

30. Assume your team of five people to develop a data base to assess testing methods for listeria in ice cream has worked together before. You expect that this team will be a high-performing team from the beginning, but you find two team members seem to disagree as to the vison of the project. You meet with these two people and find one of them really dislikes being on the project. You decide to ask your sponsor if this person can be reassigned, and another person added to the team. The sponsor agrees. This means—

    a. Initial establishment of roles and responsibilities change so the new person assumes the previous team member's work
    b. Your team is in the forming stage
    c. You should conduct an off-site team-building retreat
    d. You should meet with each team member and establish specific performance goals

31. Your organization is adopting a project-based approach to business, which has been difficult. Although project teams have been created, they are little more than a collection of functional and technical experts who focus on their specialties. You are managing the company's most important project. As you begin this project, you must place a high priority on—

    a. Guiding resource selection
    b. Identifying the resources needed to finish the project on time
    c. Determining the best way to communicate status to the CEO
    d. Establishing firm project requirements

32. Your company just learned it won a competitive proposal from your government to design a better safety system for food products in your country since there are not enough resources to inspect the imported food. You were listed in your company's proposal as the project manager since you have your PMP. You now are establishing your project team. You should consider—

    a. Schedule baseline
    b. Pre-assignments
    c. Resource calendars
    d. Problem solving

33. Now that your project to develop the next generation prostate cancer drug is under way, you have gotten to know your team members and the skills and competencies they have. You now are in a position to assess their strengths and any areas of improvement. You decided one approach to use with more junior level team members was—

    a. Mentoring
    b. Team performance assessments
    c. Motivation
    d. Communications technology

34. Work performance reports are especially helpful in the Manage Team process because—

    a. They assess team performance to take corrective or preventive action as needed
    b. They assist in determining future team resource requirements
    c. They assess the degree of the project manager's authority
    d. They document and monitor who is responsible for resolving key issues and when resolution is needed

35. Although most organizations conduct individual work performance reviews and appraisals, most of our work is done through teams. You decided since you have the luxury of a team of six people who are full-time to conduct an individual assessment but to concentrate on a team performance assessment. An example of an indicator you plan to use is—

    a. Build trust
    b. Share information
    c. Open communications
    d. Problem solving

36. You are working to estimate the activity resources you need for your new passenger train that your company will develop, own, and operate. It is to be able to travel at the speed of light, be quiet, and to be luxurious with no breakdowns in service. The goal is to test its appeal along a busy interstate highway system with stops only in three major cities. As you do so, you determined it would be helpful in addition to determining the needed resources to prepare a(n)—

    a. RACI
    b. RAM
    c. RBS
    d. OBS

37. A key benefit of the Plan Resource Management process is to determine the project organization chart. Therefore, an important first step is to—

    a. Create the WBS and let it determine the project organizational structure
    b. Review the project management plan
    c. Review the project charter
    d. Develop a project schedule, including a top-down flowchart, and identify the functional areas to perform each task

38. Assume you decide to prepare a role-responsibility-authority form as you work on your project resource management plan. This form—

    a. Provides a visual representation of resource allocation
    b. Provides detailed descriptions of team member responsibilities
    c. Sets priorities as to when scarce resources should be scheduled
    d. Shows whether the team is collocated or virtual

39. Assume you are estimating the activity resources you will need to make dirigibles a safe and fast way to travel. For this project, you will need a number of specialists as well as generalists to communicate externally as to the value of using a dirigible and its safety record. You also will need materials and equipment. As you work to estimate the resources, you believe it would be useful to consider—

    a. Resource calendars
    b. Parametric estimates
    c. Team preferences
    d. Resource requirements

40. Determining the method and the timing of releasing team members should be included in the—

    a. Staff acquisition plan
    b. Resource management plan
    c. Staffing management plan
    d. Project training plan

Answer Sheet

1.	a	b	c	d		21.	a	b	c	d
2.	a	b	c	d		22.	a	b	c	d
3.	a	b	c	d		23.	a	b	c	d
4.	a	b	c	d		24.	a	b	c	d
5.	a	b	c	d		25.	a	b	c	d
6.	a	b	c	d		26.	a	b	c	d
7.	a	b	c	d		27.	a	b	c	d
8.	a	b	c	d		28.	a	b	c	d
9.	a	b	c	d		29.	a	b	c	d
10.	a	b	c	d		30.	a	b	c	d
11.	a	b	c	d		31.	a	b	c	d
12.	a	b	c	d		32.	a	b	c	d
13.	a	b	c	d		33.	a	b	c	d
14.	a	b	c	d		34.	a	b	c	d
15.	a	b	c	d		35.	a	b	c	d
16.	a	b	c	d		36.	a	b	c	d
17.	a	b	c	d		37.	a	b	c	d
18.	a	b	c	d		38.	a	b	c	d
19.	a	b	c	d		39.	a	b	c	d
20.	a	b	c	d		40.	a	b	c	d

Answer Key

1. b. Model professional and ethical behavior

   The project manager is both the leader and manager of the project team. The project manager invests effort in acquiring team members, managing them, motivating them to do their best work, and empowering the team. The best approach is to have a participative team atmosphere from the start. One key area of influence for the project manager is to be aware of and ascribe to professional and ethical behavior and encourage the team to do the same. If the organization lacks its own code of conduct, PMI's is one to follow. [Planning]

   PMI®, *PMBOK® Guide*, 2017,
   PMI® *PMP Examination Content Outline*, 2015, Planning, 6, Task 5

2. c. Help team members work together effectively

   Team building is an interpersonal and team skill in the Develop Project Team process. It involves conducting activities that enhance the team's social relations and foster a collaborative and cooperative working relationship. Team building may consist of five minutes on an agenda of a status meeting to an off-site facilitated event. The objective of team building is the answer to this question. Although team building is a necessity in the early stage of a project, it is not a one-time activity and is done continuously during the project [Executing]

   PMI®, *PMBOK® Guide*, 2017, 341
   PMI® *PMP Examination Content Outline*, 2015, Executing, 8, Task 2

3. c. Use agreements

   Agreements and contracts are common in construction projects. Agreements are an input to the Control Resources process. These agreements are the basis for all resources external to the organization as they define when new, unplanned resources are required or when issues arise with current resources. They also define the terms and conditions and describe how long the resources will be required and how to release them when they are no longer needed. The key words in the question are leased resources. As the project manager, it is your responsibility to follow the terms in the lease and return the resources as quickly as possible, so the lease does not continue too long or is forgotten. [Monitoring and Controlling]

   PMI®, *PMBOK® Guide*, 2017, 355
   PMI® *PMP Examination Content Outline*, 2015, Monitoring and Controlling, 9, Task 1

4. d. Cultural differences

   People from different cultures may react in different ways from your culture in term of recognition and rewards. The purpose of recognition and rewards is rewarding desirable behavior. For rewards to be desired, they should satisfy a need the team member values. Reward decisions are made throughout the project both formally and informally. They are a tool and technique in the Develop Project Team process since it involves recognizing and rewarding desirable behavior. [Executing]

   PMI®, *PMBOK® Guide*, 2017, 341
   PMI® *PMP Examination Content Outline*, 2015, Executing, 8, Task 2

5. d. Colocation

   Colocation is the placement of or many or all of the most active team members in the same physical location to enhance their ability to perform as a team. It can be temporary, such as in a critical time for the project, or permanent. It may have a team meeting room, a location for project information, and other conveniences. It enhances increased communication and a sense of community among the team members. It is a tool and technique in the Develop Project Team process. [Executing]

   PMI®, *PMBOK® Guide*, 2017, 340
   PMI® *PMP Examination Content Outline*, 2017, Executing, 8, Task 2

6. b. Enhance the use of technology

   Although virtual teams have been used for decades, now they are the norm and not the exception. They enable greater access to experts, and if managed correctly and with team members in different geographic areas throughout the world, a 24-hour project can be managed. While it can be difficult to create an on-line environment for the team to share project files, see conversation threads for issues to be resolve, and keep a team calendar, the project manager can use technology to create a sense of belonging to the team among its members. [Executing]

   PMI®, *PMBOK® Guide*, 2017, 340
   PMI® *PMP Examination Content Outline*, 2015, Executing, 8, Task 2

7. b. Conflicts over the relative priorities of different projects in competition for limited resources

   When a finite group of resources must be distributed across multiple projects, conflicts in work assignments will occur. Other sources of conflicts are scheduling and personal work styles. Conflict management is a tool and technique in the Manage Project Team process. Successful conflict management results in greater productivity and more positive working relationships. The success of project managers in managing project teams often depends on their ability to resolve conflicts since conflict is inevitable in the project environment. [Executing]

   PMI®, *PMBOK® Guide*, 2017, 348–349
   PMI® *PMP Examination Content Outline*, 2015, Executing, 8, Task 2

8. c. During the execution phase

   The Manage Team process emphasizes leadership as an interpersonal team skill. It is important throughout the project. However, in Manage Project Team process, it focuses on communicating the vision of the project and then motivating and inspiring others to achieve high performance, ensuring success. Successful projects require strong leadership skills. Leadership involves wide-ranging skills, abilities, and actions, and different leadership styles are needed for each project or team. [Executing]

   PMI®, *PMBOK® Guide*, 2017, 350
   PMI® *PMP Examination Content Outline*, 2015, Executing, 8, Task 2

9. b. McGregor's Theory X

   McGregor observed two types of managers and classified them by their perceptions of workers. Theory X managers thought that workers were lazy, needed to be watched and supervised closely, and were irresponsible. Theory Y managers thought that, given the correct conditions, workers could be trusted to seek responsibility and work hard at their jobs. Ideally, the Theory X approach is not followed, but instead a flexible leadership style is used that adapts to the changes in the team's maturity throughout the project life cycle. Organizational theory is a tool and technique in Plan Resource Management. It provides information as to how people, teams, and organizational units behave. Effective use of the techniques by the leaders in this area can shorten the amount of time, cost, and effort to create the Plan Resource Management process and its outputs. Many of the leaders in this area will recommend a flexible leadership style that can adapt easily to the constant changes faced by project manager. The organization's culture and structure are other important considerations. [Planning]

   PMI®, *PMBOK® Guide*, 2017, 318
   PMI® *PMP Examination Content Outline*, 2015, Planning, 6, Task 5

10. a. Roles and responsibilities section

    The roles and responsibilities section of this plan states the role or the function to be assumed or assigned to someone on the project; the authority to apply project resources, make decisions, sign approvals, accept deliverables, and influence others to execute the work; the responsibilities or assigned duties and work for project team members to complete project activities; and competencies or the skill or capacity to complete project activities. If team members lack the required competencies, overall performance may be jeopardized. If there are mismatches then other strategies are needed such as training, reassignment, hiring someone else, schedule changes, or scope changes. [Planning]

    PMI®, *PMBOK® Guide*, 2017, 318–319
    PMI® *PMP Examination Content Outline,* 2015, Planning, 6, Task 5

11. a. Enhanced project performance

    The purpose of the Develop Team process is enhanced project performance. It is accomplished by improving competencies, team member interaction, and the overall team environment. The other answers are benefits of this process. [Executing]

    PMI®, *PMBOK® Guide*, 2017, 336
    PMI® *PMP Examination Content Outline,* 2015, Executing, 8, Task 2

12. c. Establish a team charter

    The team charter is an output of Plan Resource Management. Its purpose is to state team values, agreements, and operating guidelines. It also can contain communication strategies, decision-making criteria and processes to follow, describe a process to resolve conflicts, and provide meeting guidelines. The best approach is to have the team prepare the charter and have each team member sign it to indicate agreement with it and to show commitment to it. It should be evaluated periodically to see if it is working as expected, and if not, it should be changed and updated. [Planning]

    PMI®, *PMBOK® Guide*, 2017, 319–320
    PMI® *PMP Examination Content Outline,* 2015, Planning, 6, Task 5

13. d. Team directory

    The project team directory is part of project team assignments, an output from the Acquire Project Team process. The team directory incudes names, which can be part of the project management plan such as in the project organization and schedule. [Executing]

    PMI®, *PMBOK® Guide*, 2017, 334
    PMI® *PMP Examination Content Outline,* 2015, Executing, 8, Task 1

14. b. Chart

   All of the answers are a type of chart that can be used. Also included in this category of data representation as a tool and technique is the work breakdown structure and text-oriented formats. Most project managers use a hierarchical, matrix, or text-based format. Some use all methods. A hierarchical method, for example, is useful for high--level roles, while a text-based format is helpful to document detailed responsibilities. [Planning]

   PMI®, *PMBOK® Guide*, 2017, 316–317
   PMI® *PMP Examination Content Outline*, 2015, Planning, 6, Task 5

15. a. Problem solving

   In the Control Resources process, problem solving is a tool and technique to use in this type of situation as in this question. It uses a set of tools the project manager can use to solve problems that can arise in this process. Steps in problem solving are to: identify the problem, define it, investigate, analyze it, solve it, and check the solution. [Monitoring and Controlling]

   PMI®, *PMBOK® Guide*, 2017, 356
   PMI® *PMP Examination Content Outline*, 2015, Monitoring and Controlling, 9, Task 1

16. b. Authority section

   Authority refers to the right to apply project resources, make decisions, accept deliverables, sign approvals, and influence others to want to do the project work. Examples include selecting methods to complete activities, quality acceptance criteria, and responding to variances in the project. The individual authority of each team member should match their individual responsibilities; when this is done it is easier for the team members to do their work. This is documented in the authority section in the resource management plan. [Planning]

   PMI®, *PMBOK® Guide*, 2017, 318
   PMI® *PMP Examination Content Outline*, 2015, Planning, 6, Task 5

17. d. Resource management plan

   The resource management plan is the key output of Plan Resource Management. It provides guidance as to how resources should be acquired, categorized, allocated, managed, and controlled. While it has a number of sections, the one that is pertinent to this question is acquiring resources in which guidance as to how to acquire team and physical resources is provided. [Planning]

   PMI®, *PMBOK® Guide*, 2017, 318
   PMI® *PMP Examination Content Outline*, 2015, Planning, 6, Task 5

18. c. Resolve it in a timely way

In the Develop Project Team process, interpersonal and team skills are a tool and technique. Within this category is conflict management. It is necessary when there are conflicts that the project manager resolves them as quickly as possible. While there are a variety of approaches to use, the quicker it is resolved and communicated to the affected people, the less likely it is to spread and become a larger conflict affecting more people. [Executing]

PMI®, *PMBOK® Guide*, 2017, 341
PMI® 1*PMP Examination Content Outline*, 2015, Executing, 8, Task 2

19. b. Include a human resource representative in your RAM chart

As you prepare your resource management plan, charts are a tool and technique to use. In this situation, using a variant of the resource assignment matrix, you can use a RACI (responsible, accountable, consult, and inform) chart and show various activities and who is responsible in each category in the four areas. In it, a human resources representative can be shown as someone to consult in certain situations. The assigned resources in this chart can be people or groups. [Planning]

PMI®, *PMBOK® Guide*, 2017, 317
PMI® *PMP Examination Content Outline*, 2015, Planning, 6, Task 5

20. a. Is in the storming stage

Team development leads to improved team performance, which ultimately results in improved project performance. According to the model developed by Bruce Tuckman, your team is one that is in the storming stage, and you believe this off-site event can help them pass through it quickly, so they can reach the norming stage. Storming is when the team begins to address the work of the project, has technical decisions to make, and needs to follow the project management approach. It is easy for the team to be one that is counterproductive in this stage if they are not collaborative or open to different ideas. [Executing]

PMI®, *PMBOK® Guide*, 2017, 338
PMI® *PMP Examination Content Outline*, 2015, Executing, 8, Task 2

21. c. Collaborating

    Collaborating or problem solving is an effective technique for managing conflict when a project is too important to be compromised. It involves incorporating multiple insights and viewpoints from people with different perspectives and offers a good opportunity to learn from others. It provides a long-term resolution. It requires a cooperative attitude and open dialogue that typically leads to consensus and commitment. [Executing]

    PMI®, *PMBOK® Guide*, 2017, 349
    PMI® *PMP Examination Content Outline,* 2015, Executing, 8, Task 2

22. a. The organizational structure of your company

    Enterprise environmental factors can influence the Develop Resource Management process. The organizational structure of the performing organization determines whether the project manager's role is a strong one (as in a strong matrix) or a weak one (as in a weak matrix). Other examples of enterprise environmental factors are the organization's culture, geographic dispersion of team members, existing human resources competencies and availability, and marketplace conditions. [Planning]

    PMI®, *PMBOK® Guide*, 2017, 315
    PMI® *PMP Examination Content Outline,* 2015, Planning, 6, Task 5

23. c. Improving feelings of trust and agreement

    Developing a project team improves people skills, technical competencies, and the overall team environment and project performance. Improving feelings of trust and agreement among team members can raise morale, lessen conflicts, and increase team work. [Executing]

    PMI®, *PMBOK® Guide*, 2013, 282
    PMI® *PMP Examination Content Outline,* 2015, Executing, 7, Task 2

24. d. Provide individual and team assessments

As a tool and technique in the Develop Team process, individual and team assessments are used to give the project manager and the team insight into areas of strength and areas in need of improvement. They can help the project manager assess team member preferences, aspirations, how they process and organize information, make decisions, and interact with people. Tools such as attitude surveys, structured interviews, ability tests, and focus groups can be used. By using these types of tools, the objective is to improve understanding, trust, commitment, and communications among the team, which can facilitate a more productive team for the remainder of the project. [Executing]

PMI®, *PMBOK® Guide*, 2017, 309
PMI® *PMP Examination Content Outline,* 2015, Executing, 8, Task 2

25. c. Work with your team to prepare a responsibility assignment matrix

The responsibility assignment matrix (RAM) defines project roles and responsibilities in terms of work packages and activities. On large projects, a high-level RAM can define responsibilities of a project team, group, or unit for each WBS component. Lower-level RAMs are used in the team to show roles, responsibilities, and levels of authority for specific activities. The matrix shows all activities associated with one person, and all people associated with an activity. It is a tool and technique in data representation in Plan Resource Management. [Planning]

PMI®, *PMBOK® Guide*, 2017, 317
PMI® *PMP Examination Content Outline,* 2015, Planning, 6, Task 5

26. d. Meetings

The key words in the question is the project manager seeks opinions of others. Of the various estimating approaches available in the four answers, meetings are the perfect choice. By using meetings, the project manager can hold meetings with functional managers, the project sponsor, selected stakeholders, and others in the organization. The objective is to estimate needed resources per activity, level of effort, skills required, and the materials needed. [Planning]

PMI®, *PMBOK® Guide*, 2017, 325
PMI® *PMP Examination Content Outline,* 2015, Planning, 6, Task 5

27. b. Build on the meetings you held with the key people and use influencing skills

    Resources are scarce in most organizations. The best approach to follow is to effectively negotiate and influence others who are in a position to provide the resources needed for the team and the project. If the resources cannot be acquired and if external procurement is not available, the project is in trouble from the start and may need to be put on hold until resources can be provided. [Executing]

    PMI®, *PMBOK® Guide*, 2017, 330
    PMI® *PMP Examination Content Outline,* 2015, Executing, 8, Task 1

28. b. Economic factors

    During the process of Acquire Resources, the project manager may be required to assign alternative resources. Examples of constraints are economic factors or previous assignments to other projects. The project manager then may need to assign people who perhaps have different competencies or costs, provided there is no violation of legal, regulatory, mandatory, or other specific criteria. [Executing]

    PMI®, *PMBOK® Guide*, 2017, 330
    PMI® *PMP Examination Content Outline,* 2015, Executing, 8, Task 1

29. c. Attitude

    Attitude is used to determine one's ability to work with others as a cohesive team. Other selection criteria for team resources are experience, knowledge, skills, and international factors. A multi-criteria approach is useful as the criteria are developed and used to rate or score potential resources, such as those internal to the organization or external to it. The criteria also can be weighted according to their importance, and values can be changed for different types of resources. [Executing]

    PMI®, *PMBOK® Guide*, 2017, 323
    PMI® *PMP Examination Content Outline,* 2015, Executing, 8, Task 1

30. b. Your team is in the forming stage

    Even if team members have worked together previously, it cannot be assumed they will immediately be a high-performing team. According to the Tuchman team development model, all teams have a forming stage in which the team meets and learns about the project and their formal roles and responsibilities. At this time, team members are independent and not as open in this phase. This model further goes on to the storming stage, norming stage, performing stage, and adjourning stage. However, even if the team is in the norming or performing stage, if a team member leaves and/or a new person joins, the team will revert back to the forming stage. [Executing]

    PMI®, *PMBOK® Guide*, 2017, 338
    PMI® *PMP Examination Content Outline,* 2015, Executing, 8, Task 2

31. a. Guiding resource selection

    The purpose of the Acquire Resources process is to obtain team members, facilities, equipment, supplies, and any other resources needed to do the work. The benefit of this process is the answer to the question as well as assigning them to activities. Resources can be internal or external to the organization. The project manager and the team may or may not have a direct control over resource selection because of collective bargaining as an example. Other examples are use of contractor personnel, a matrix environment, reporting relationships, or other reasons. [Executing]

    PMI®, *PMBOK® Guide*, 2017, 328–329
    PMI® *PMP Examination Content Outline,* 2015, Executing, 8, Task 1

32. b. Pre-assignments

    Pre-assignments are a tool and technique in the Acquire Resources process. While you were listed in the proposal in this question as the project manager, in acquiring team resources, or physical resources, other resources also may have been listed in the proposal as well. Pre-assignment also includes resources identified in the Develop Project Charter process or in other processes before the resource management plan was completed. [Planning]

    PMBOK® Guide, 2017, 333
    PMI® *PMP Examination Content Outline,* 2015, Planning, 6, Task 5

33. a. Mentoring

   Mentoring is useful as a training approach in which a person on the team is paired with someone with more experience, hopefully from elsewhere in the organization. The objective is to work one-on-one to address issues the mentee may have and determine possible approaches to resolve them or to address other areas of concern. For it to be effective, a mentoring plan should be prepared, time needs to be set aside for mentoring sessions, the mentor should receive training in how to approach it, and the mentee must want to do it. While it is useful with junior-level employees with a more seasoned person in the organization, it also can be used in reverse. For example, assume a senior-level team member is having problems adapting to a new enterprise resource planning software system, he or she could be mentored by a more junior-level person who may have been involved in its design and implementation. Mentoring is not coaching, as coaching is an approach that is directing, while mentoring is something one may express an interest in pursuing. Answer b is an output of this process. [Executing]

   PMBOK® Guide, 2017, 342
   PMI® *PMP Examination Content Outline*, 2015, Executing, 8, Task 2

34. b. They assist in determining future team requirements

   Work performance reports are an input in the Manage Team process. These reports are physical or electronic representations used to determine decisions, actions, or awareness. In the Manage Team process, they also include results from schedule, cost, and quality control and scope validation. The information from the performance reports and any forecasts are used to determine future resource requirements, along with rewards and recognition. They also may be used to update the resource management plan. [Executing]

   PMI®, *PMBOK® Guide*, 2017, 347
   PMI® *PMP Examination Content Outline*, 2015, Executing, 8, Task 2

35. b. Share information

Team performance assessments are an output of the Develop Team process. The objective is to assess the project team's effectiveness. By doing so, effective team development strategies and activities should increase the team's performance, which then increases the potential to meet the project objectives. Indicators such as sharing information can be used to change from a knowledge is power approach to a knowing sharing is power approach and shows the team members are working together to not have duplicative work under way. If information and experiences are shared openly, it can help team members improve project performance. These team assessments also assist the project manager in identifying specific training activities or changes to improve team performance and to identify any resources needed to achieve and implement the assessment recommendations. [Executing]

PMI®, *PMBOK® Guide*, 2017, 343
PMI® *PMP Examination Content Outline*, 2015, Executing, 8, Task 2

36. c. RBS

While the WBS shows how project deliverables are broken down into work packages to show high-level areas of responsibility, the organizational breakdown structure (OBS) is arranged according to the organization's departments, units, or teams with the work packages listed under each one as appropriate. The resource breakdown structure (RBS) is a hierarchical list of resources related by category and resource type used to facilitate and control the work of the project. Each descending lower level represents an increasing detailed description of the resource unit that is small enough to be used in conjunction with the WBS to enable the work to be planned, monitored, and controlled. It contains all resources, not solely people. Examples of categories are labor, materials, equipment, and supplies. Resource types include skill level, grade level, and required certifications. In the Estimate Activity Resources process, it is used to acquire and monitor resources. [Planning]

PMI®, *PMBOK® Guide*, 2017, 326–327
PMI® *PMP Examination Content Outline*, 2015, Planning, 6, Task 5

37. c. Review the project charter

    The first input in the Plan Resource Management plan process is to review the project charter as it describes the high-level project description and requirements. It also identifies the key list of stakeholders, summary milestones, and preapproved financial resources, which may influence project resource management. In preparing the resource plan, it is useful to determine and identify an approach where you can ensure sufficient resources are available for the project in the planning phase to complete the project. Resources are scarce, and project managers often compete for them. The signed charter gives the project manager the authority to assign resources to the project and also has the available financial resources. This means the project is important in terms of the overall project portfolio as it has been chartered. The benefit of this processes is it identifies the approach and level of management of the project needed for managing the resources considering the project's size and complexity. [Planning]

    PMI®, *PMBOK® Guide*, 2017, 312–314
    PMI® *PMP Examination Content Outline*, 2015, Planning, 6, Task 5

38. b. Provides detailed descriptions of team member responsibilities

    In preparing the resource management plan, data representation is a tool and technique to use. Various charts are part of this tool and technique to document and communicate team member roles and responsibilities. While many consider the RAM or RACI charts, not to be overlooked is a text-oriented format, which may be a position description or a role-responsibility-authority form. These documents provide information such as responsibilities, authority, competencies, and qualifications. They may be in an outline form and often then are templates for future projects. [Planning]

    PMI®, *PMBOK® Guide*, 2017, 317
    PMI® *PMP Examination Content Outline*, 2015, Planning, 6, Task 5

39. a. Resource calendars

Resource calendars are an input to the Estimate Activity Resources process. They are useful since they identify the working days, shifts, start and end times of normal business hours, weekends, and public holidays when each resource is available. Information on the resources, including more than just people, are available during a planned activity period is used to estimate resource use. These calendars also show how long the team and physical resources will be available. Information about resource use may be at the activity or project level. The calendars can include items such as resource experience, skill level, and geographic locations. [Planning]

PMI®, *PMBOK® Guide*, 2017, 323
PMI® *PMP Examination Content Outline*, 2015, Planning, 6, Task 5

40. b. Resource management plan

The resource management plan is the key output of Plan Resource Management. This plan provides guidance on categorizing project resources, allocating them, managing them, and releasing them. While there are several sections in this plan, one specifically addresses project team resource management. It is here guidance on how project team resources should be defined, staffed, managed, and eventually released. [Planning]

PMI®, *PMBOK® Guide*, 2017, 318–319
PMI® *PMP Examination Content Outline*, 2015, Planning, 6, Task 5

Project Communications Management

Practice Questions

INSTRUCTIONS: Note the most suitable answer for each multiple-choice question in the appropriate space on the answer sheet.

1. The communications management process goes beyond distributing relevant information as it—

   a. Focuses on establishing working relationships and standard formats for global communication among stakeholders
   b. Enables an effective and efficient information flow between the project team and its stakeholders
   c. Establishes individual and group responsibilities and accountabilities for communication management
   d. Protects confidential and sensitive information

2. One purpose of the communications management plan is to provide information about the—

   a. Methods that will be used to convey information
   b. Methods that will be used for releasing team members from the project when they are no longer needed
   c. Project organization and stakeholder responsibility relationships
   d. Experience and skill levels of each team member

3. Project managers for international projects should recognize key issues in cross-cultural settings and place special emphasis on—

   a. Establishing a performance reporting system
   b. Using effective communication planning
   c. Establishing and following a production schedule for information distribution to avoid responding to requests for information between scheduled communications
   d. Using translation services for formal, written project reports

4. You are managing a project with team members located at customer sites on three different continents. As you plan communications with your stakeholders, you should review—

   a. Stakeholder management plan
   b. Stakeholder register
   c. Communications model
   d. Communications channels

5. Having worked previously on projects as a team member, you are pleased to now be the project manager to develop a new process to ensure that software projects in your IT Department are considered a success and are not late or over budget. Many of your team members are new to the organization. You are working to prepare your communications management plan for your project. You want to describe communications as a process between two parties, so you decide to use—

   a. The communications strategy
   b. Sender-receiver model
   c. Cross-cultural communications model
   d. Interactive communications model

6. As a project manager, you try to use active listening skills to help understand another person's frame of reference. In following this approach, you should—

   a. Mimic the content of the message
   b. Probe, then evaluate the content
   c. Remove barriers that affect comprehension
   d. Rephrase the content and reflect the feeling

7. Statements of organizational policies and philosophies, position descriptions, and constraints are examples of—

   a. Formal communication
   b. Lateral communication
   c. External communication
   d. Horizontal communication

8. You have decided to organize a study group of other project managers in your organization to help prepare for the PMP® exam. What type of communication activity are you employing in your efforts to organize this group?

   a. Horizontal
   b. Vertical
   c. Official
   d. External

9. Your company CEO just sent you an e-mail asking you to make a presentation on your project, which has been in progress for 18 months, to over 50 identified internal and external stakeholders. You have been conducting such presentations and holding meetings regularly on this important project. You should begin by—

   a. Defining the audience
   b. Providing background information
   c. Deciding on the general form of the presentation
   d. Circulating issues to be discussed

10. You are responsible for a project in your organization that has multiple internal customers. Because many people in your organization are interested in this project, you realize the importance of—

   a. Conducting a stakeholder analysis to assess information needs
   b. Performing communications planning early
   c. Determining the communications requirements of the customers
   d. Having an expert on communications management and customer relationship management on your team

11. Project managers spend a great deal of time communicating with the team, the stakeholders, the client, and the sponsor. One can easily see the challenges involved, especially if one team member must communicate a technical concept to another team member in a different country. While there are different communications models to use, in this situation the first step is to—

   a. Encode the message
   b. Decode the message
   c. Determine the feedback loops
   d. Determine the emotional state

12. On your project, scope changes, constraints, assumptions, integration and interface requirements, and overlapping roles and responsibilities pose communications challenges. The presence of communication barriers is most likely to lead to—

    a. Reduced productivity
    b. Increased hostility
    c. Low morale
    d. Increased conflict

13. The most common communication problem that occurs during negotiation is that—

    a. Each side may misinterpret what the other side has said
    b. Each side may give up on the other side
    c. One side may try to confuse the other side
    d. One side may be too busy thinking about what to say next to hear what is being said

14. You finally have been appointed project manager for a major company project. One of your first activities as project manager will be to create the communications management plan. One of your goals is to use real time communications. This means as you share information among your stakeholders, you are using—

    a. Interactive communications
    b. Small group conversations
    c. Pull communications
    d. Push communications

15. As an output of Plan Communications, it may be necessary to update the project documents, which include the—

    a. Stakeholder register
    b. Corporate policies, procedures, and processes
    c. Knowledge management system
    d. Stakeholder management plan

16. Sample attributes of a communications management plan include which one of the following?

    a. Roles
    b. Responsibilities
    c. Ethics
    d. Authority

17. Feedback is a communication skill in Manage Communications. In it one receives information about his or her skill in—

    a. Active listening
    b. Negotiations
    c. Influencing
    d. Presentations

18. The key benefit of the Monitor Communications process is—

    a. Sharing best practices with other project teams in the organization with lessons learned
    b. Ensuring the information needs of stakeholders are met
    c. Ensuring an optimal information flow among communication participants
    d. Providing stakeholders with information about resolved issues, approved status, and project status

19. The issue log is useful in Monitor Communications because it—

    a. Provides the project's history
    b. Includes the project's risk register
    c. Organizes and summarizes information gathered
    d. Serves as an information management system for communications management

20. As head of the PMO, you will receive performance reports for all major projects. You decided to set a guideline for project managers as performance reporting should—

    a. Collect work performance information on the status of deliverables
    b. Provide earned value data for project forecasting
    c. Provide information at an appropriate level for each stakeholder
    d. Focus on cost and schedule variances rather than scope, resources, quality, and risks

21. You are the project manager to devise and implement a strategy for your University to have greater recognition in the project management field. It has classes in all of the ten knowledge areas in project management, but few outside of its local area know about it. As you prepare your strategy, you decide it would be useful to form online communities with alumni, which are—

    a. Included in your project management plan
    b. Described in your communications plan
    c. Part of your project management information system
    d. The responsibility of your PMO

22. Communication is important when setting and managing expectations with the stakeholders. In preparing your communications management plan you realize—

    a. You need to determine your stakeholders' information needs.
    b. The choice of media is a key concern.
    c. Flexibility in communications methods.
    d. Electronic communications methods are a goal.

23. In person-to-person communication, messages are sent on verbal levels and nonverbal levels simultaneously. To communicate effectively, the project manager should be—

    a. Aware of the communication styles of the other parties
    b. Aware of their own preferred communication style
    c. Able to apply flexible style to meet audience requirements
    d. Aware of non-verbal communications

24. As an output from Monitor Communications, it may be necessary to update the—

    a. Risk register
    b. Issue log
    c. Corporate policies, procedures, and processes
    d. Knowledge management system

25. In project communications, the first step in a written communication is to—

    a. Determine the appropriate writing style
    b. Gather thoughts or ideas
    c. Develop a logical sequence of the topics to be addressed
    d. Establish the basic purpose of the message

26. A communications management plan includes which one of the following sample contents?

    a. Issues
    b. Escalation processes
    c. Dimensions
    d. Project assumptions and constraints

27. Your organization has decided to use project management for all of its endeavors. It has established a Center of Excellence for Project Management to support the movement into management by projects and has appointed you as its director. You have decided to focus on the methods used to transfer information, which then will be part of project communications management plans. In deciding the preferred approach, you need to determine—

    a. The type of information that is used the most
    b. The urgency of the need for information
    c. Reporting formats
    d. The preferred communications model

28. You hear repeatedly that your client finds the format of the progress reports you submit to be one that does not meet their requirements. The client representatives complain they do not understand data in the reports. You know this is the case, since soon after you submit a report, the client representative calls you immediately to express concerns. Obviously, the communications requirements analysis you conducted is flawed. It also could be the communications methods you are using may not be effective. This situation shows it could have been avoided by—

    a. Meeting with the client at the start of the project to discuss the types of reports to prepare
    b. Using risk management techniques to identify client issues
    c. Hiring an expert report writer to prepare standard reports
    d. Engaging in communications planning

29. Assume on your project you have identified 250 stakeholders located in three continents and of these 250, you have determined that 200 of them will be actively involved and interested in your project. Therefore, as you determine an appropriate communication method, your best approach is—

    a. Elaborate status reports
    b. Simple status reports
    c. Knowledge repositories
    d. E-mails

30. You want to ensure that the information you collect showing project progress and status is meaningful to stakeholders. However, while you have been managing projects for 15 years, you are new to this company. You decide as you prepare your communications management plan to—

    a. Review the organization chart
    b. Determine the disciplines, departments and specialists who will be involved with your project
    c. Focus on the logistical requirements to see the people involved at various locations
    d. Determine the communications requirements

31. Work performance information is an output of which process?

    a. Manage risks
    b. Manage communications
    c. Monitor communications
    d. Report performance

32. Assume you want to optimize the work performance reports you will use to Manage Communications. You should do so by—

    a. Determining the most appropriate choice of communications media
    b. Setting different communications techniques for different stakeholder groups
    c. Ensuring the information is consistent with regulations and standards
    d. Circulating them to stakeholders as defined in the communications management plan

33. Planned and ad hoc reports and presentations are a useful update to organizational process assets during which following process?

    a. Plan Communications Management
    b. Distribute Information
    c. Manage Communications
    d. Monitor Communications

34. An example of a useful communications skill used in Manage Communications is—

    a. Political awareness
    b. Setting and managing expectations
    c. Nonverbal
    d. Influencing

35. Changes in the communications management and stakeholder engagement plans should trigger changes to the—

   a. Project management plan
   b. Communications strategy
   c. Lessons learned register
   d. Change control process

36. One way to determine how to best update and communicate project performance and respond to stakeholder information requests is to—

   a. Review the effectiveness of the communications management plan
   b. Set up a portal
   c. Hold meetings
   d. Distribute performance reports

37. The purpose of work performance data in Monitor Communications is to—

   a. Obtain data on organizational communications requirements
   b. Review distributed information
   c. Review stakeholder communication strategies
   d. Evaluate deliverable status

38. Because communications planning often is linked tightly with enterprise environmental factors, you should recognize that—

   a. The project's organizational culture has an effect on the project's communications requirements.
   b. Standardized guidelines, work instructions, and performance measurement criteria are key items to consider.
   c. Procedures for approving and issuing work authorizations should be taken into consideration.
   d. Criteria and guidelines to tailor standard processes to the specific needs of the project should be stated explicitly.

39. You are working on a project with 15 stakeholders. Each of your 10 team members have a secret security clearance, and as the project manager, you have a top-secret security clearance as you work for a government contractor. As you prepare your communications management plan, you need to determine—

   a. Whether the 15 stakeholders also require a secret security clearance
   b. Social media policies
   c. Ways to distribute project information given this environment
   d. The impact of these clearances to the project environment

40. You have surveyed the stakeholders in your stakeholder register to help determine their communications requirements. The majority of the respondents, 95%, want information distributed directly to them. This situation shows—

    a. You should use a web portal.
    b. Push communications are preferred.
    c. Pull communications are preferred.
    d. A knowledge transfer system is required.

Answer Sheet

1.	a	b	c	d		21.	a	b	c	d
2.	a	b	c	d		22.	a	b	c	d
3.	a	b	c	d		23.	a	b	c	d
4.	a	b	c	d		24.	a	b	c	d
5.	a	b	c	d		25.	a	b	c	d
6.	a	b	c	d		26.	a	b	c	d
7.	a	b	c	d		27.	a	b	c	d
8.	a	b	c	d		28.	a	b	c	d
9.	a	b	c	d		29.	a	b	c	d
10.	a	b	c	d		30.	a	b	c	d
11.	a	b	c	d		31.	a	b	c	d
12.	a	b	c	d		32.	a	b	c	d
13.	a	b	c	d		33.	a	b	c	d
14.	a	b	c	d		34.	a	b	c	d
15.	a	b	c	d		35.	a	b	c	d
16.	a	b	c	d		36.	a	b	c	d
17.	a	b	c	d		37.	a	b	c	d
18.	a	b	c	d		38.	a	b	c	d
19.	a	b	c	d		39.	a	b	c	d
20.	a	b	c	d		40.	a	b	c	d

Answer Key

1. b. Enables an effective and efficient information flow between the project team and its stakeholders

   The answer is the key benefit of the Manage Communications process. In addition, the Manage Communications process involves collection, creation, distribution, storage, retrieval, management, monitoring, and disposition of information about the project. [Executing]

   PMI®, *PMBOK® Guide*, 2017, 379
   PMI® *PMP Examination Content Outline*, 2015, Executing, 8, Task 6

2. a. Methods that will be used to convey information

   These methods or technologies can include memos, e-mails, press releases, and social media. They are one of several items to include in this plan. This plan, an output of the Plan Communications Management process, describes how project communications will be planned, structured, implemented, and monitored. [Planning]

   PMI®, *PMBOK® Guide*, 2017, 377
   PMI® *PMP Examination Content Outline*, 2015, Planning, 6, Task 6

3. b. Using effective communications planning

   Cultural awareness recognizes the differences between people, groups, and organizations. The project manager and team adapt the communications strategy to these differences. The objective is to minimize any misunderstandings and miscommunications from cultural differences within the project's stakeholders. By being aware of the cultural differences and having sensitivity to them, the project manager can use this information to effectively plan communications for the project. Further, it is useful to include a glossary of common terminology in the communications management plan. [Planning]

   PMI®, *PMBOK® Guide*, 2017, 376–377
   PMI® *PMP Examination Content Outline*, 2015, Planning, 6, Task 6

4. b. Stakeholder register

   The stakeholder register and the requirements documentation are two project documents that are an input to the Plan Communications Management process. The stakeholder register contains identification information, assessment information, and stakeholder classification. It is used to help plan communications activities with stakeholders. [Planning]

   PMI®, *PMBOK® Guide*, 2017, 368, 514
   PMI® *PMP Examination Content Outline*, 2015, Planning, 6, Task 6

5. b. Sender-receiver model

Communications models are a tool and technique in Plan Communications Management. In the sender-receiver model, communications are a process with two parties - the sender and receiver. Its purpose is to ensure the message is delivered rather than focusing on whether it was understood. It has three steps. First, the message is encoded into symbols, which can include test, sound, or another medium. Second, the message is transmitted through a communication channel. Third, the data are decoded or translated by the receiver to a form that is useful. The key words in this question are the person is new project manager so he or she would probably use this simpler approach. [Planning]

PMI®, *PMBOK® Guide, 2017,* 371
PMI® *PMP Examination Content Outline,* 2017, Planning, 6, Task 6

6. c. Remove barriers that affect comprehension

Active listening is one of the interpersonal and team skills used in Manage Communications. While the tendency is to work to improve one's presentation skills, less attention is given to improving one's listening skills. With active listening the focus is on acknowledging, clarifying, and confirming information and to understand and then remove any barriers to communications. It Is necessary to remove these barriers as they may affect comprehension. It also means being engaged with the speaker and summarizing conversations for an attentive information exchange. [Executing]

PMI®, *PMBOK® Guide,* 2017, 363, 366
PMI® *PMP Examination Content Outline,* 2015, Executing, 8, Task 6

7. a. Formal communication

Formal communication provides direction and control for project team members and other employees. They consist of regular and ad hoc meetings, meeting agendas, minutes, stakeholder presentations, and briefings. Formal communications are examples of organizational process assets used in Communications Management.

PMI®, *PMBOK® Guide,* 2017, 361, 369
PMI® *PMP Examination Content Outline,* 2015, Planning, 6, Task 6

8. a. Horizontal

   Communication activities have many potential dimensions to consider in exchanging information between the sender and the receiver. Horizontal communication occurs between or among peers of the project manager or the team, that is, across, rather than up and down, the organization. [Executing]

   PMI®, *PMBOK® Guide*, 2017, 361
   PMI® *PMP Examination Content Outline*, 2015, Executing, 8, Task 6

9. b. Providing background information

   Presentations are held regularly on projects to update and communicate project information and to respond to requests from stakeholders for the information. They are a communication skill used in Manage Communications. In this situation the project manager has been conducting these types of presentations regularly. Even so, he or she should begin with a brief overview of background information about the project since some of the stakeholders who are attending have multiple projects in which they are involved to some degree. This background information can be used to support decision making. The presentation can serve as a progress report to stakeholders, can provide general information about the project and its objectives such that the project may raise the profile of the work under way, and can provide specific information to increase understanding and support of the project's objectives. [Executing]

   PMI®, *PMBOK® Guide*, 2017, 384
   PMI® *PMP Examination Content Outline*, 2015, Executing, 8, Task 6

10. b. Performing communications planning early

    On most projects, communications planning should be performed very early such as during stakeholder identification and when the project management plan is prepared. This approach then allows appropriate resources, such as time and budget, to be allocated to communications activities. The results of Plan Communications Management should be reviewed regularly and updated as needed. The goal is to ensure the communications management plan recognizes the diverse stakeholders involved with the project as the stakeholder community often changes. It also can be reviewed at the beginning of a new project phase. [Planning]

    PMI®, *PMBOK® Guide*, 2017, 367
    PMI® *PMP Examination Content Outline*, 2015, Planning, 6, Task 6

11. d. Determine the emotional state

The key words in this question is that technical information is being transmitted to team members in different countries. This means the cross-cultural communications model is useful to ensure the meaning of the message is understood. People from different cultures communicate in different ways. The cross-cultural communications model emphasizes the idea that the message and how it is sent are influenced by the sender's current emotional state. Knowledge transfer, background, personality, and biases can influence how the message is received and interpreted. [Planning]

PMI®, *PMBOK® Guide*, 2017, 373
PMI® *PMP Examination Content Outline*, 2015, Planning, 6, Task 6

12. d. Increased conflict

Barriers to communication lead to a poor flow of information. Accordingly, messages are misinterpreted by recipients, thereby creating different perceptions, understanding, and frames of reference. Left unchecked, poor communication increases conflict among project stakeholders, which causes the other problems listed to arise. Then, the project manager must work actively to resolve conflicts so disruptive impacts are prevented. Conflict management is included as a tool and technique under interpersonal and team skills in Manage Communications. [Executing]

PMI®, *PMBOK® Guide*, 2017, 386
PMI® *PMP Examination Content Outline*, 2015, Executing, 8, Task 6

13. a. Each side may misinterpret what the other side has said

Effective communication is the key to successful negotiation, a key communication skill that the project manager and the team should enhance. Misunderstanding is the most common communication problem. A project manager should listen actively, acknowledge what is being said, and speak for a purpose. It is essential to listen attentively and communicate articulately. Negotiation is necessary to achieve mutually acceptable agreements between both parties and to reduce approval or delays in decision making. Negotiation is an integral part of project management and if done well increases the probability of project success. [Executing]

PMI®, *PMBOK® Guide*, 2017, 363
PMI® *PMP Examination Content Outline*, 2015, Executing, 8, Task 6

14. a. Interactive communications

The types of communications are a tool and technique in Plan Communications Management. While there are several approaches to use, the key words in this question are real time. Interactive communication involves exchanging information in real time between two or more parties. It uses meetings, phone calls, instant messages, social media, and video conferencing. [Planning]

PMI®, *PMBOK® Guide*, 2017, 374
PMI® *PMP Examination Content Outline*, 2015, Planning, 6, Task 6

15. a. Stakeholder register

In the Plan Communications Management process the two documents that may be updated are the project schedule and the stakeholder register. The stakeholder register is an input to this process but may be changed as the communications management plan is prepared based on communications requirements analysis. The stakeholder register then is updated to reflect planned communications to project stakeholders. [Planning]

PMI®, *PMBOK® Guide*, 2017, 378
PMI® *PMP Examination Content Outline*, 2015, Planning, 6, Task 6

16. b. Responsibilities

Among other things, the communications management plan should identify the person responsible for communicating the information and the person responsible for authorizing the release of any confidential information. [Planning]

PMI®, *PMBOK® Guide*, 2017, 377
PMI® *PMP Examination Content Outline*, 2015, Planning, 6, Task 6

17. b. Negotiations

Negotiation if done well increases the probability of project success and involves conferring with others of shared or opposed interests with a view toward compromise. In Manage Communications, it is useful to receive feedback concerning one's negotiating skills. Other examples where feedback is useful are coaching and mentoring. Feedback involves reactions to communications, a deliverable, or a situation. It supports interaction between the project manager, the team, and other stakeholders. [Executing]

PMI®, *PMBOK® Guide*, 2017, 384
PMI® *PMP Examination Content Outline*, 2015, Executing, 8, Task 6

18. c. Ensuring an optimal information flow among all communication participants

    While Monitor Communications confirm the communication needs of project stakeholders are met, the key benefit is to ensure an optimal information flow among all communication participants at any moment in time. This process determines if the planned communications artifacts and activities have had the desired effect of increasing or maintaining stakeholder support for the project. This process is performed throughout the project. This process can trigger the iteration of the Plan Communications Management and Manage Communications processes to improve communications effectiveness by possibly amending the communications management plan. [Monitoring and Controlling]

    PMI®, *PMBOK® Guide*, 2017, 388–389
    PMI® *PMP Examination Content Outline*, 2015, Monitoring and Controlling, 9, Task 1

19. a. Provides the project's history

    The issue log is a project document that is an input to Monitor Communications. It is helpful because it provides the project history, serves as a record of stakeholder engagement issues, and describes how these issues were resolved. [Monitoring and Controlling]

    PMI®, *PMBOK® Guide*, 2017, 390
    PMI® *PMP Examination Content Outline*, 2015, Monitoring and Controlling, 9, Task 1

20. c. Provide information at an appropriate level for each stakeholder

    Project reporting is a tool and technique in Manage Communications. It involves collecting and distributing project information. These reports range from simple status reports to more elaborate reports. The emphasis is to ensure reports are adapted to provide the information in a format and at the right level of detail for each stakeholder. Reports may be prepared on a regular or an exception basis. In Manage Communications ad hoc reports, presentations, and other project communications are made available. [Executing]

    PMI®, *PMBOK® Guide*, 2017, 385
    PMI® *PMP Examination Content Outline*, 2015, Executing, 8, Task 6

21. c. Part of your project management information system

    The PMIS is a tool and technique in Manage Communications. It ensures stakeholders can easily retrieve needed information in a timely way. Project information is managed and distributed in three ways, one of which is social media management. Social media consists of websites, web publishing, blogs, and applications. The purpose is to engage with stakeholders and also provide the opportunity to form online communities. [Executing]

    PMI®, *PMBOK® Guide*, 2017, 385
    PMI® *PMP Examination Content Outline*, 2015, Executing, 8, Task 6

22. a. You need to determine your stakeholders' information needs.

    Communications is considered one of the single most powerful indicators of project success or failure. Communications requirements analysis is a tool and technique in Plan Communications Management. By analyzing communications requirements, you can determine the information needs of your stakeholders. By doing this analysis, it then leads to the ability to combine the type and format of information needs and an analysis of the value of the information. A number different sources can be used as you analyze communications requirements. [Planning]

    PMI®, *PMBOK® Guide*, 2017, 369
    PMI® *PMP Examination Content Outline*, 2015, Planning, 6, Task 6

23. a. Aware of the communication styles of the other parties

    A communication styles assessment is an interpersonal and team skill, which is a tool and technique in Plan Communications Management. This assessment is helpful to assess the communication styles and identify the preferred method, format, and content stakeholders prefer for planned communications. This assessment often is used when some stakeholders have been identified as ones who are not supporters of the project, and it may follow a stakeholder engagement assessment. [Planning]

    PMI®, *PMBOK® Guide*, 2017, 375
    PMI® *PMP Examination Content Outline*, 2015, Planning, 6, Task 6

24. b. Issue log

    Monitor Communications often entails the need to update project documents, including the issue log. This log may require updates as new issues may be raised. Other project documents that may require updates are the lessons learned register and the stakeholder register. [Monitoring and Controlling]

    PMI®, *PMBOK® Guide*, 2017, 393
    PMI® *PMP Examination Content Outline*, 2015, Monitoring and Controlling, 9, Task 1

25. a. Determine the appropriate writing style

    In Manage Communications, it goes further than distributing relevant information as its objective is to ensure the information being communicated has been gathered and received by the intended people. It also provides the opportunity for stakeholders to make requests for further information. While there are a variety of methods to distribute information, written communication is used frequently. It is easy for the receiver to misinterpret it, but the objective is for the receiver to understand it and use it as desired. Writing style then is a consideration and involves the use of active voice or passive voice, sentence structure, and word choice. [Executing]

    PMI®, *PMBOK® Guide*, 2017, 381
    PMI® *PMP Examination Content Outline*, 2015, Executing, 6, Task 6

26. b. Escalation processes

    Numerous items, including escalation processes, are part of the communications management plan. Business issues may arise that cannot be resolved at a lower staff level. During such a time, an escalation process is required, and it should show time frames and the names of people in the management chain who will work to resolve these issues. [Planning]

    PMI®, *PMBOK® Guide*, 2017, 377
    PMI® *PMP Examination Content Outline*, 2015, Planning, 6, Task 6

27. b. The urgency of the need for information

    Communications technology is a tool and technique in Plan Communications. The methods used to convey information to stakeholders will vary. Conversations, meetings, written documents, data bases, social media, and web sites are used, among others. Determining the urgency of the need for information will affect the choice of technology. The urgency, frequency, and format will vary from project to project as there is no one approach that can be used. It even can vary within different phases of the project. [Panning]

    PMI®, *PMBOK® Guide*, 2017, 370
    PMI® *PMP Examination Content Outline,* 2015, Planning, 6, Task 6

28. a. Meeting with the client at the start of the project to discuss the types of reports to prepare

    The communications management plan is prepared during Plan Communications Management. The plan should include a description of the information to be distributed such as format, content, level of detail, as well as conventions and definitions to be used. It also can include templates and guidelines for meetings and e-mails. In preparing this plan, and especially in this scenario in the question, meetings are a useful tool and technique. They can be used, among other things, to determine the most appropriate way to update and communicate project information and to respond to stakeholders. By having client representatives as part of such meetings, the problems in this scenario may have been prevented. [Planning]

    PMI®, *PMBOK® Guide*, 2017, 376
    PMI® *PMP Examination Content Outline,* 2015, Planning, 6, Task 6

29. c. Knowledge repositories

    Knowledge repositories, along with Intranet sites, e-learning, and lessons learned data bases, are examples of methods of pull communications. Pull communications are a type of communication method as a tool and technique in Plan Communications Management. They are used for large volumes of information or for large audiences and require recipients to access communication content at their own discretion, recognizing the need to have security provisions in place as needed. [Planning]

    PMI®, *PMBOK® Guide*, 2017, 374
    PMI® *PMP Examination Content Outline,* 2015, Planning, 6, Task 6

30. d. Determine the communications requirements

    The project team must conduct an analysis of stakeholder communications requirements to ensure that stakeholders are receiving the information required to participate in the project. For example, stakeholders typically require performance reports for information purposes. Such information requirements should be included in the communications management plan. Conducting a communications requirements analysis to determine the stakeholders' information needs is a tool and technique in the Plan Communications process. The other answers are considerations as you conduct the communications requirements analysis. [Planning]

    PMI®, *PMBOK® Guide*, 2017, 369–370
    PMI® *PMP Examination Content Outline,* 2015, Planning, 5, Task 6

31. c. Monitor communications

    Work performance information, an output of Monitor Communications, provides information as to how project communications were performed by comparing the communications that were used against those that were planned. It considers feedback on communications to assess effectiveness. [Monitoring and Controlling]

    PMI®, *PMBOK® Guide*, 2017, 392
    PMI® *PMP Examination Content Outline,* 2015, Monitoring and Controlling, 9, Task 1

32. d. Circulating them to stakeholders as defined in the communications management plan

    Work performance reports are distributed to stakeholders through processes in the communications management plan. They can include status and progress reports. They can contain earned value data, trends and forecasts, reserve burndown charts, defect histograms, contract data, and risk summaries. These reports are an input to Manage Communications. [Executing]

    PMI®, *PMBOK® Guide*, 2017, 382
    PMI® *PMP Examination Content Outline,* 2015, Executing, 8, Task 6

33. c. Manage Communications

Planned and ad hoc reports and presentations are a useful organizational process asset to update from Manage Communications. The other group of organizational process assets to update are project records, such as correspondence, memos, meeting minutes, and other documents used on the project. [Executing]

PMI®, *PMBOK® Guide*, 2017, 388
PMI® *PMP Examination Content Outline*, 2015, Executing, 8, Task 6

34. c. Nonverbal

Communications skills include nonverbal communications. Examples are body language, as it transmits messages through gestures, tone of voice, and facial expressions. Mirroring and eye contact also are important in this category. The project manager and the team members should be aware as to how their use of nonverbal communications is noted by others. [Executing]

PMI®, *PMBOK® Guide*, 2017, 384
PMI® *PMP Examination Content Outline*, 2015, Executing, 8, Task 6

35. a. Project management plan

As an output of the Monitor Communications process, there may be changes to the communications management plan to update it to make it more effective. Also, there may be changes to the stakeholder engagement plan to reflect their actual situation, communications needs, and their importance. Changes to these plans then trigger changes to the project management plan. These changes follow the organization's change control process. [Monitoring and Controlling]

PMI®, *PMBOK® Guide*, 2017, 393
PMI® *PMP Examination Content Outline*, 2015, Monitoring and Controlling, 9, Task 2

36. c. Hold meetings

Meetings are a tool and technique in Monitor Communications. They can be face to face or online and in different locations and may include not only the project team but also suppliers, vendors, and other stakeholders. It is important to hold these meetings for decision making, to respond to requests from stakeholders, and to have discussions with suppliers, vendors, and other stakeholders. [Monitoring and Controlling]

PMI®, *PMBOK® Guide*, 2017, 392
PMI® *PMP Examination Content Outline*, 2015, Monitoring and Controlling, 9, Task 1

37. b. Review distributed information

    Work performance data are an input in Monitor Communications. These data organize and summarize the types and quantities of communications. [Monitoring and Controlling]

    PMI®, *PMBOK® Guide*, 2017, 390
    PMI® *PMP Examination Content Outline*, 2015, Monitoring and Controlling, 9, Task 1

38. a. The project's organizational culture has an effect on the project's communications requirements.

    Enterprise environmental factors undoubtedly will influence the project's success and must be considered because communication must be adapted to the project environment. They are an input to the Plan Communications Management Plan process. The organization's culture, political climate, and governance framework are examples of enterprise environmental factors to consider. [Planning]

    PMI®, *PMBOK® Guide*, 2017
    PMI® *PMP Examination Content Outline*, 2015, Planning, 6, Task 6

39. b. Social media policies

    Sensitivity and confidentiality of information is part of communications technology, an input to Plan Communications. It is necessary to consider if the information is sensitive or confidential and whether additional security measures are needed. Also, with the increase in social media, employees must ensure appropriate behavior, security, and the protection of proprietary information are followed. [Planning]

    PMI®, *PMBOK® Guide*, 2017, 371
    PMI® *PMP Examination Content Outline*, 2015, Planning, 6, Task 6

40. b. Push communications are preferred.

    Communication methods are an input to Plan Communications Management. When push communications are used, information is sent directly to specific recipients who want to receive it. This approach shows the information is distributed, but it does not show it was actually received or was understood. It includes letters, memos, reports, emails, faxes, voice mails, blogs, and press releases. [Planning]

    PMI®, *PMBOK® Guide*, 2017, 374
    PMI® *PMP Examination Content Outline*, 2015, Planning, 6, Task 6

Project Risk Management

Practice Questions

INSTRUCTIONS: Note the most suitable answer for each multiple-choice question in the appropriate space on the answer sheet.

1. As the project manager, you have the option of proposing one of three systems to a client: a full-feature system that not only satisfies the minimum requirements but also offers numerous special functions (the "Mercedes"); a system that meets the client's minimum requirements (the "Smart Car"); and a system that satisfies the minimum requirements plus has a few extra features (the "Toyota"). The on-time records and associated profits and losses are depicted on the below decision tree. What is the expected monetary value of the "Toyota" system?

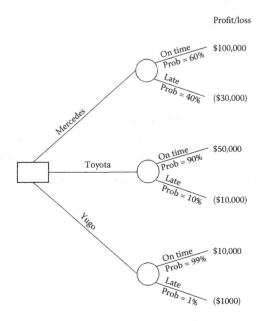

a. $9,900
b. $44,000
c. $45,000
d. $48,000

2. A risk response strategy that can be used for both threats and opportunities is—

   a. Share
   b. Avoid
   c. Accept
   d. Transfer

3. The risk urgency is a tool and technique used for—

   a. Plan Risk Responses
   b. Identify Risks
   c. Perform Qualitative Risk Analysis
   d. Perform Quantitative Risk Analysis

4. Projects are particularly susceptible to risk because—

   a. People assume there will be problems
   b. There is uncertainty in all projects
   c. Project management tools are generally unavailable at the project team level
   d. There are never enough resources to do the job

5. As project manager, you have assembled the team to prepare a risk management plan for your project. It is to describe how risk management will be handled on your project. Within this plan, there is a—

   a. OBS
   b. WBS
   c. RBS
   d. CCB

6. You are working on identifying possible risks to your project to develop a nutritional supplement. You want to develop a comprehensive list of risks that can be addressed later through qualitative and quantitative risk analysis. A data gathering technique used to identify risks is—

   a. Documentation reviews
   b. Probability and impact analysis
   c. Prompt lists
   d. Brainstorming

7. Assume you have prepared your risk management plan, identified possible risks, analyzed them, and determined possible responses if they occur. Now it is time to Implement Risk Responses. The benefit of this process is to—

    a. Present a sequence of decision choices to decision makers

    b. Ensure agreed-upon risk response are executed as planned

    c. Ensure actions to implement risks are only taken from people who are on the team

    d. Help take into account the attitudes of the decision makers toward risk

8. You are working Monitor Risks on your manufacturing project. You are fortunate to have a subject matter expert available with a specialty in risk management to assist you and your team. As you work to monitor these risks, you want to make sure your team and stakeholders are aware of—

    a. Any unplanned response to a negative risk event

    b. A plan of action to follow when something unexpected occurs

    c. The level of overall project risk

    d. A proactive, planned method of responding to risks

9. Most statistical simulations of budgets, schedules, and resource allocations use which one of the following approaches?

    a. PERT

    b. Decision-tree analysis

    c. Present value analysis

    d. Monte Carlo analysis

10. Assume you are using probability and impact analysis. In the below example, if the odds of completing activities 1, 2, and 3 on time are 50 percent, 50 percent, and 50 percent, what are the chances of starting activity 4 on day 6?

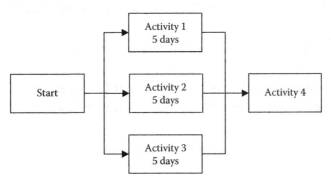

    a. 10 percent

    b. 13 percent

    c. 40 percent

    d. 50 percent

11. A project health check identified a risk that your project would not be completed on time. You and your team are working to prevent this from occurring as you work to Implement Risk Responses. You and your team know risk management is continuous on projects, and new risks will occur. When they do occur analysis is needed, and a response strategy is determined. Your goal is to see if the strategy is executed. As you do so, you should use—

    a. Negotiations
    b. Influencing skills
    c. Schedule network diagram and duration estimates
    d. Probability/impact risk rating matrix

12. You are developing radio frequency (RF) technology that will improve overnight package delivery to one hour. You realize that the duration, cost, and resource requirements for a number of activities are uncertain. You want a model to represent individual project risk and other sources of uncertainty. This shows that for your next step you plan to—

    a. Use a probability distribution
    b. Conduct a sensitivity analysis
    c. Structure a decision analysis as a decision tree
    d. Determine the strategy for risk response

13. You are determining the most effective risk response to a risk one of your team members identified. After analyzing it and including a risk management expert in the analysis process, it is evident it is a risk where the threat is such that it is outside of your authority as the project manager to determine a response. In this situation, you need to—

    a. Focus on changing the project management plan to eliminate the threat
    b. Escalate it to the next highest level
    c. Solicit the advice by the head of the Enterprise Project Management Office
    d. Change the project objective that is in jeopardy

14. If the probability of event 1 is 80 percent and of event 2 is 70 percent and they are independent events, how likely is it that both events will occur?

    a. 6 percent
    b. 15 percent
    c. 24 percent
    d. 56 percent

15. The project scope statement should be used in the Identify Risk process because it—

    a. Identifies project assumptions
    b. Identifies all the work that must be done and, therefore, includes all the risks on the project
    c. Helps to organize all the work that must be done on the project
    d. Contains information on risks from prior projects

16. Your project team has identified all the risks on the project and has categorized them as high, medium, and low. As you work to Monitor Risks, you want to consider all three categories as a low risk can change quickly to a medium or high risk given changes that the project may encounter. Therefore, you plan to—

    a. Consider the threat level
    b. Hold meetings
    c. Set up a watch list
    d. Review work performance information

17. A contingency reserve is used for—

    a. Risks for which the accept response is selected
    b. Risks that are not identified at the outset of the project but are known before they occur
    c. Risks that use a contingent response strategy
    d. Risk for which the avoid strategy is used

18. The simplest form of quantitative risk analysis and modeling techniques is—

    a. Probability analysis
    b. Sensitivity analysis
    c. Delphi technique
    d. Utility theory

19. If a business venture has a 60-percent chance to earn $2 million and a 20-percent chance to lose $1.5 million, what is the expected monetary value of the venture?

    a. −$50,000
    b. $300,000
    c. $500,000
    d. $900,000

20. You are managing the construction of a highly sophisticated data center in Southeast Asia. Although this location offers significant economic advantages, the threat of typhoons has caused you to create a backup plan to operate in Manila in case the center is flooded. This plan is an example of what type of risk response?

    a. Contingent response
    b. Mitigation
    c. Accept
    d. Transfer

21. A recent earned value analysis shows that your project is 20 percent complete, the CPI is 0.67, and the SPI is 0.87. In this situation, you should—

    a. Perform additional resource planning, add resources, and use overtime as needed to accomplish the same amount of budgeted work
    b. Rebaseline the schedule, then use Monte Carlo analysis
    c. Conduct a risk response audit to help control risk
    d. Forecast potential deviation of the project at completion from cost and schedule targets

22. The purpose of a numeric scale in risk management is to—

    a. Avoid high-impact risks
    b. Assign a relative value to the impact on project objectives if the risk in question occurs
    c. Rank order risks in terms of very low, low, moderate, high, and very high
    d. Test project assumptions

23. Risk score measures the—

    a. Variability of the estimate
    b. Product of the probability and impact of the risk
    c. Range of schedule and cost outcomes
    d. Reduced monetary value of the risk event

24. In Monitor Risks the project manager, the team, and key stakeholders need to be aware of the level of risk exposure; new, changing, or outdated risks; and changes in the level of overall project risk. Information from project execution then is used to determine whether to—

    a. Conduct a risk audit
    b. Engage in additional risk response planning
    c. Modify contingency reserves
    d. Conduct a risk review

25. The primary advantage of using decision tree analysis in project risk management is that it—

    a. Considers the attitude of the decision maker toward risk
    b. Forces consideration of the probability of each outcome
    c. Helps to identify and postulate risk scenarios for the project
    d. Supports the selection of the best approach of several alternatives

26. Your project is using complex, unproven technology. Your team conducted a brainstorming session to identify risks. Poor allocation of project resources was the number one risk. This risk was placed on the risk register, which included at this point a—

    a. Watch list
    b. Potential risk response
    c. Known unknown
    d. List of other risks requiring additional analysis

27. When managing current projects, it is important to use lessons learned from previous projects to improve the organization's project management process. Therefore, in the Identify Risks process one can review the—

    a. Process flow charts
    b. Checklists
    c. WBS
    d. Root-cause analysis

28. Risk mitigation involves—

    a. Using performance and payment bonds
    b. Eliminating a specific threat by eliminating the cause
    c. Avoiding the schedule risk inherent in the project
    d. Reducing the probability and/or impact of an adverse risk event to an acceptable threshold

29. Risk management is not a one-time process and transcends project management. Risks are first discussed as part of the—

    a. Project plan
    b. Project charter
    c. Business case
    d. Enterprise environmental factors

30. Two key inputs to the Perform Quantitative Risk Analysis process are the—

    a. WBS and milestone list
    b. Scope management plan and process improvement plan
    c. Schedule baseline and cost baseline
    d. Procurement management plan and quality baseline

31. Often in preparing the risk management plan meetings are held. The key purpose of these meetings is to—

    a. Involve those involved in executing the plan
    b. Enable anyone in the organization to participate
    c. Define how best to conduct risk management activities
    d. Ensure the key stakeholders are invited and participate

32. As you work to Identify Risk, an item to review is—

    a. Activity estimates
    b. Duration estimates
    c. Communications management plan
    d. Procurement management plan

33. A risk audit is a useful tool and technique in Monitor Risks because it—

    a. Focuses on implementing agreed-upon risk responses
    b. Considers the effectiveness of the risk management process
    c. Enables risk reassessments to be performed
    d. Serves as a risk review

34. Accurate and unbiased data are essential for Perform Qualitative Risk Analysis. The best approach to follow is to use—

    a. Data quality assessment
    b. Project assumptions testing
    c. Sensitivity analysis
    d. Influence diagrams

35. Assigning more talented resources to the project to reduce time to completion or to provide better quality than originally planned are examples of which one of the following strategies?

    a. Enhance
    b. Exploit
    c. Share
    d. Contingent response

36. You are striving to determine whether to accept a risk or transfer it to a contractor. You also may decide to just avoid the risk. You can—

    a. Collect more data using questionnaires
    b. Use influencing skills to gather stakeholders' opinions
    c. Conduct an assumptions analysis
    d. Use decision making

37. Assume you remain unclear about what to do about a possible risk that may affect the overall project. You and your team have several strategies you can use, but before selecting one, you decide to—

    a. Conduct an alternatives analysis
    b. Establishing a management reserve to cover unplanned expenditures
    c. Review the stakeholder engagement plan
    d. Determine needed adjustments to make during the implementation phase of a project

38. Assume that you are working on a new product for your firm. Your CEO learned that a competitor was about to launch a new product that has similar features to those of your project. The competitor plans to launch the product on September 1. It is now March 1. Your schedule called for you to launch your product on December 1. Your CEO now has now mandated that you fast track your project, so you can launch your product on August 1. This fast track schedule is an example of an—

    a. Unknown risk
    b. A risk taken to achieve a reward
    c. A risk that impacts strategic goals
    d. A passive avoidance strategy

39. As head of the project management office, you need to focus on those items where risk responses can lead to better project outcomes. One way to help you make these decisions is to—

    a. Use interviews
    b. Assess trends in Perform Quantitative Risk Analysis results
    c. Prioritize risks and conditions
    d. Assess trends in Perform Qualitative Risk Analysis results

40. You are the project manager for the construction of an incinerator to burn refuse. Local residents and environmental groups are opposed to this project. Management agrees to move this project to a different location. This is an example of which one of the following risk responses?

    a. Passive acceptance
    b. Active acceptance
    c. Mitigation
    d. Avoidance

Answer Sheet

1.	a	b	c	d
2.	a	b	c	d
3.	a	b	c	d
4.	a	b	c	d
5.	a	b	c	d
6.	a	b	c	d
7.	a	b	c	d
8.	a	b	c	d
9.	a	b	c	d
10.	a	b	c	d
11.	a	b	c	d
12.	a	b	c	d
13.	a	b	c	d
14.	a	b	c	d
15.	a	b	c	d
16.	a	b	c	d
17.	a	b	c	d
18.	a	b	c	d
19.	a	b	c	d
20.	a	b	c	d

21.	a	b	c	d
22.	a	b	c	d
23.	a	b	c	d
24.	a	b	c	d
25.	a	b	c	d
26.	a	b	c	d
27.	a	b	c	d
28.	a	b	c	d
29.	a	b	c	d
30.	a	b	c	d
31.	a	b	c	d
32.	a	b	c	d
33.	a	b	c	d
34.	a	b	c	d
35.	a	b	c	d
36.	a	b	c	d
37.	a	b	c	d
38.	a	b	c	d
39.	a	b	c	d
40.	a	b	c	d

Answer Key

1. b. $44,000

$$EMV_{\text{Toyota}} = (\$50{,}000 \times 90\%) + (-\$10{,}000 \times 10\%)$$
$$= \$45{,}000 + (-\$1{,}000)$$
$$= \$44{,}000$$

Expected monetary value analysis (EVM) is a statistical concept that calculates the average outcome when the future scenarios may not happen, considered analysis under uncertainty. The EVM opportunities are generally evaluated as positive values, as in this example, while threats are expressed as negative values. They are used in decision tree analysis, which is a tool and technique in quantitative risk analysis. The purpose is to support selecting the best of several alternative courses of action. The decision tree shows branches that represent different decisions or events. Each one has costs and related individual project risks. The end points of the branches show the outcome from following a path, which is either positive or negative. [Planning]

PMI®, *PMBOK® Guide*, 2017, 435
PMI® *PMP Examination Content Outline*, 2015, Planning, 6, Task 10

2. c. Accept

Risk exists on every project, and it is unrealistic to think it can be eliminated completely. There are certain risks that simply must be accepted because we cannot control whether or not they will occur (for example, an earthquake). Acceptance is a strategy for dealing with risk that can be used for both threats and opportunities. Using it, the project team decides to acknowledge the risk and not take any action. It is appropriate for low-priority threats or opportunities or if it is not possible or cost effective to address a threat opportunity in another way. [Planning]

PMI®, *PMBOK® Guide*, 2017, 443–444
PMI® *PMP Examination Content Outline*, 2015, Planning, 6, Task 10

3. c. Perform Qualitative Risk Analysis

Risks that may happen in the near-term need urgent attention. Risk urgency is part of many considerations in assessments of other risk parameters, a data analysis tool and technique in Perform Qualitative Risk Analysis. In assessing other parameters, the project team considers other characteristics of risk, such as urgency, in addition to probability or impact as the team prioritizes individual project risks for additional analysis or action. The purpose of the risk urgency assessment is to identify those risks that have a high likelihood of happening sooner rather than later. It assesses the time when a response to the risk is implemented to be effective. High urgency is shown by a short period of time. [Planning]

PMI®, *PMBOK® Guide*, 2017, 423–424
PMI® *PMP Examination Content Outline*, 2015, Planning, 6, Task 10

4. b. There is uncertainty in all projects

Every project has uncertainty associated with it because a project by its definition is a temporary endeavor undertaken to create a unique product, service, or result. Projects have varying degrees of complexity. As projects work to deliver benefits, they also are working with constraints and assumptions as they respond to expectations of stakeholders that may conflict and change. The goal is to identify and manage risks that are not covered in other project management processes. If risks are not managed, they may cause the project to deviate from the plan and fail to meet their objective. Effective risk management contributes to successful projects. [Planning]

PMI®, *PMBOK® Guide*, 2017, 397
PMI® *PMP Examination Content Outline*, 2015, Planning, 6, Task 10

5. c. RBS

The risk breakdown structure (RBS) is a way to group individual project risks. It is a hierarchically organized depiction of potential sources of risk. It helps the team consider a number of sources where project risks may arise. It is useful in categorizing and identifying risks. Different RBS structures are appropriate for different types of projects, or the organization can use a previously prepared framework that can be a simple list of categories, or it may be structured into a RBS. It is described in the risk categories section of the risk management plan. [Planning]

PMI®, *PMBOK® Guide*, 2017, 405–406
PMI® *PMP Examination Content Outline*, 2015, Planning, 6, Task 10

6. d. Brainstorming

Brainstorming is a frequently used data gathering technique for identifying risk. It enables the project team to develop a list of potential risks relatively quickly. Project team members, often with invited experts, possibly under the direction of a facilitator, participate in the session. Traditional brainstorming sessions are free form, or other approaches may be used. Categories of risks, such as in a RBS, can be used as a framework. The identified risks should be described as brainstorming may result in ideas that may not be completely formed. It is part of several data gathering approaches as data gathering is a tool and technique in Identify Risks. [Planning]

PMI®, *PMBOK® Guide*, 2013, 324
PMI® *PMP Examination Content Outline,* 2015, Planning, 5, Task 10

7. b. Ensure agreed-upon risk response are executed as planned

The purpose of the Implement Risk Response process is to implement agreed-upon risk responses. The key benefit is it ensures these responses are executed as planned such that it addresses overall project risk exposure, minimizes individual potential threats, and maximizes project opportunities. [Executing]

PMI®, *PMBOK® Guide*, 2017, 449
PMI® *PMP Examination Content Outline,* 2015, Executing, 8, Task 2

8. c. The level of overall project risk

The Monitor Risk process involves monitoring the agreed-upon risk management plans, tracking and identifying new risks, and evaluating the effectiveness of risk responses throughout project. You want your team and other stakeholders to be aware of the current level of risk exposure and to recognize project work that should be continuously monitored for new risks, outdated risks, and changes in the level of overall project risk. The level of overall project risk is important to see if it has changed. [Monitoring and Controlling]

PMI®, *PMBOK® Guide*, 2017, 454
PMI® *PMP Examination Content Outline,* 2015, Monitoring and Controlling, 9, Task 4

9. d. Monte Carlo analysis

A project simulation uses a model that translates the specified detailed uncertainties into their potential outcomes on project objectives and is a data analysis tool and technique in Perform Quantitative Risk Analysis, Simulations are typically performed using Monte Carlo analysis in which a project model is computed many times. The input values for cost risk simulation use project cost estimates, and for simulations of schedule risks, the network diagram and duration estimates are used. [Planning]

PMI®, *PMBOK® Guide*, 2017, 433
PMI® *PMP Examination Content Outline*, 2015, Planning, 6, Task 10

10. b. 13 percent

$$\text{Probability (starting activity 4 on day 6)} = (0.5)^3$$
$$= 0.125 \text{ or } 13\%$$

Such an approach helps to prioritize risks for further quantitative analysis and to plan risk responses based on their risk rating. The rating is based on the assessed probability and impact for each individual risk and the impact on one or more project objectives. It uses the definitions of probability and impact from the risk management plan. The probability and impact matrix is a data analysis tool and technique in the Perform Qualitative Risk Analysis process. [Planning]

PMI®, *PMBOK® Guide*, 2017, 425
PMI® *PMP Examination Content Outline*, 2015, Planning, 6, Task 10

11. b. Influencing skills

In the Implement Risk Responses process, interpersonal skills are a tool and technique. Within it, influencing is suggested. Some risk response actions may be owned by a stakeholder who is not part of the project team. These stakeholders tend to have competing demands for their time, and the same applies to project team member working on more than one project and working in a matrix structure. The project manager or the person responsible for the project's risk management activities then uses influencing to encourage risk owners to take action. [Executing]

PMI®, *PMBOK® Guide*, 2017, 451
PMI® *PMP Examination Content Outline*, 2015, Executing, 8, Task 5

12. a. Use a probability distribution

Representations of uncertainty is a tool and technique in quantitative risk analysis. With the types of uncertainties described in the question, the range of possible values is represented in a model as a probability distribution. The most common forms are triangular, normal, lognormal, beta, uniform, or discrete. Individual risks may be covered by the distribution. The best one to use should reflect the range of possible results for the activity. [Planning]

PMI®, *PMBOK® Guide*, 2017, 432
PMI® *PMP Examination Content Outline*, 2015, Planning, 6, Task 10

13. b. Escalate it to the next highest level

Escalate is a risk response strategy for both threats and opportunities. In this question, it is a threat as it is outside of the scope of this project or as in the question outside the scope of the project manager's authority. An escalated risk is managed at the program level, the portfolio level, or in another part of the organization; not at a project level. The project manager determines whom to notify and provide details of the work done thus far. Usually, threats are escalated to the level that is involved with the affected project objective for resolution if the threat would occur. These threats may be listed in the project's risk register, but the team is no longer involved with this risk. [Planning]

PMI®, *PMBOK® Guide*, 2017, 442
PMI® *PMP Examination Content Outline*, 2015, Planning, 6, Task 10

14. d. 56 percent

The likelihood is determined by multiplying the probability of event 1 by the probability of event 2. While probability investigates the likelihood that a specific risk will occur, impact investigates the possible effect on the project's objective such as schedule, quality, or performance. A probability and impact grid is a data representation tool and technique in Perform Qualitative Risk Analysis. The grid maps the probabilities of each risk occurrence and its impact on project objectives should the risk occur. The matrix can be set up to show individual risks by priority groups. The risks then can be prioritized further for additional analysis and to determine a specific response. Risk probabilities and impact are rated according to definitions in the risk management plan. [Planning]

PMI®, *PMBOK® Guide*, 2017, 425–426
PMI® *PMP Examination Content Outline*, 2015, Planning, 6, Task 10

15. a. Identifies project assumptions

    Project assumptions, which should be enumerated in the project scope statement, are areas of uncertainty, and as such are potential causes of project risk. Uncertainty of assumptions is evaluated as potential causes of project risk and include project assumptions and constraints. The scope statement, the WBS, and the WBS Dictionary are part of the scope baseline, part of the project management plan, an input to Identify Risks. [Planning]

    PMI®, *PMBOK® Guide*, 2017, 413
    PMI® *PMP Examination Content Outline*, 2015, Planning, 5, Task 10

16. b. Hold meetings

    Meetings are a tool and technique in Control Risks. They may be formal or informal but should be used to document risk response effectiveness considering both overall project risk and identified individual project risks. New risks may be identified, including secondary risks in the meeting. Lessons learned for this project and other projects in the organization may be identified. [Monitoring and Controlling]

    PMI®, *PMBOK® Guide*, 2017, 457
    PMI® *PMP Examination Content Outline*, 2015, Monitoring and Controlling, 9, Task 4

17. a. Risks for which the accept response is selected

    Risk response strategies should be selected for both individual project risks plus the overall project risk. The accept strategy is used when no proactive risk strategy is available for overall project risk, and the organization decides to continue the project even if the overall risk is outside of the agreed-upon threshold. Acceptance can be active or passive. If an active acceptance strategy is selected, an overall contingency reserve for the project including time, money, or resources is set aside should it be necessary to use it if the risk exceeds its thresholds. [Planning]

    PMI®, *PMBOK® Guide*, 2017, 446
    PMI® *PMP Examination Content Outline*, 2015, Planning, 6, Task 10

18. b. Sensitivity analysis

   Sensitivity analysis, as a quantitative risk analysis data analysis tool and technique, helps to determine which risks or other sources of uncertainty have the most potential impact on the project's outcomes. It examines the extent to which the variations in the project's objectives correlate with variations in the quantitative risk analysis model. A tornado diagram often is used to show the calculated correlation coefficient for each element in the model that can affect the project outcomes. [Planning]

   PMI®, *PMBOK® Guide*, 2017, 434
   PMI® *PMP Examination Content Outline*, 2015, Planning, 6, Task 10

19. d. $900,000

   $$EMV = (\$2M \times 60\%) + (-\$1.5M \times 20\%) = (\$1.2M) + (-\$300{,}000)$$
   $$= \$900{,}000$$

   As another example of a quantitative risk analysis and modeling technique, EMV is a statistical concept in decision tree analysis that calculates the average outcome when the future includes scenarios that may or may not happen. This is an example of an opportunity with a positive value. Its use requires a risk-neutral assumption, and it is calculated by multiplying the value of each possible outcome by the probability of occurrence and adding the products together. [Planning]

   PMI®, *PMBOK® Guide*, 2017, 435
   PMI® *PMP Examination Content Outline*, 2015, Planning, 6, Task 10

20. a. Contingent response

   The contingent response strategy is designed for use if certain events occur, as in this question - a typhoon. This response plan is only executed under pre-defined conditions when there is time to use it, such as a warning that a typhoon is coming. Other examples to consider are if a key milestone was missed or if using it results in a higher priority with a seller. Risk responses using this technique may be called contingency plans or fallback plans. [Planning]

   PMI®, *PMBOK® Guide*, 2017, 445
   PMI® *PMP Examination Content Outline*, 2015, Planning, 6, Task 10

21. d. Forecast potential deviation of the project at completion from cost and schedule targets

   Earned value is used for monitoring overall project performance against a baseline plan. Outcomes from the analysis may forecast potential deviation of the project at completion from its cost and schedule targets. Deviations from the baseline plan also may indicate the potential impact of threats or opportunities. It is a part of earned value data, an input to Monitor Risks in work performance reports. [Monitoring and Controlling]

   PMI®, *PMBOK® Guide*, 2017, 456
   PMI® *PMP Examination Content Outline*, 2015, Monitoring and Controlling, 9, Task 4

22. b. Assign a relative value to the impact on project objectives if the risk in question occurs

   You can develop relative or numeric, well-defined scales using agreed-upon definitions by the stakeholders. When using a numeric scale, each level of impact has a specific number assigned to it. Each risk then is rated on its probability and impact on an objective if it does occur. Evaluation of each risk's importance and priority is typically done by using a look-up table or a probability and impact matrix. This matrix specifies conditions of probability and impact that lead to ratings of risk such as low, moderate, or high priority. Descriptive terms or numeric value can be used. [Planning]

   PMI®, *PMBOK® Guide*, 2017, 407–408
   PMI® *PMP Examination Content Outline*, 2015, Planning, 6, Task 10

23. b. Product of the probability and impact of the risk

   Using a probability and impact matrix, the project manager and the team can rate a risk separately for an objective, and they can develop ways to determine the overall rating for each risk. Threats and opportunities can be shown in the same matrix. The risk score provides a convenient way to compare risks because comparing impacts or probabilities alone is meaningless. It helps guide risk responses. [Planning]

   PMI®, *PMBOK® Guide*, 2017, 407–408, 423
   PMI® *PMP Examination Content Outline*, 2015, Planning, 6, Task 10

24. c. Modify contingency reserves

    The purpose of the Monitor Risks process is to implement the agreed-upon risk responses, track risks, identify and analyze existing risks, and evaluate the effectiveness of project risk management. Although this process uses tools and techniques, this answer, among others, is an example as to how the Monitor Project Risks process uses performance information from execution. It may be necessary to determine whether the contingency reserves for cost or schedule need modification. [Monitoring and Controlling]

    PMI®, *PMBOK® Guide*, 2017, 453–454
    PMI® *PMP Examination Content Outline*, 2015, Monitoring and Controlling, 9, Task 4

25. d. Supports the selection of the best approach of several alternatives

    Decision tree analysis is a tool and technique in Perform Quantitative As a graphical way to bring together information, decision tree analysis quantifies the likelihood of failure or opportunities and places a value on each decision. In the decision tree, alternative branches are shown. They represent different decisions or events, each one having different costs and individual risks. The end point in the analysis represents the preferred outcome. It is evaluated through expected monetary value analysis. [Planning]

    PMI®, *PMBOK® Guide*, 2017, 435
    PMI® *PMP Examination Content Outline*, 2015, Planning, 6, Task 10

26. b. Potential risk response

    The risk register is prepared in the Identify Risks process as an output. It is a document in which the results of risk analysis and risk response planning are recorded, and as they are conducted it contains the outcomes of the other risk management processes. The identified risks are described in as much detail as is reasonable. They may use structured risk statements to distinguish risks from their causes and effects. Each risk in the register has an identification number and is described in as much detail as possible. Risk owners may be included. Potential risk responses also may be identified during the Identify Risks process, and if this is the case they will be confirmed in the Plan Risk Responses process. [Planning]

    PMI®, *PMBOK® Guide*, 2017, 417
    PMI® *PMP Examination Content Outline*, 2015, Planning, 6, Task 10

27. b. Checklists

Checklists are a tool and a technique in the Identify Risks process. Often, they are used as a reminder and include risks encountered on similar, previous projects based on historical information and knowledge. They are effective in capturing lessons learned by listing specific risks that may have occurred on other projects and may be relevant for this project. Generic risk checklists also are available. However, while it is easy to prepare a checklist and simple to use, it is impossible to make sure it covers all items relative to the project, and some items on it may not apply to the project. It should be reviewed occasionally during the project to update it with new information and to remove any obsolete data from it. [Planning]

PMI®, *PMBOK® Guide*, 2017, 414
PMI® *PMP Examination Content Outline*, 2015, Planning, 6, Task 10

28. d. Reducing the probability and/or impact of an adverse risk event to an acceptable threshold

It is often more effective to take early action to reduce probability and/or impact of a risk occurring on a project rather than attempting to repair the damage after the risk has occurred. Examples of mitigation or enhancement strategies include adopting less complex processes, conducting more tests, or choosing a more stable supplier. Other examples are replacing the project, changing its scope, modifying its priority, and adjusting delivery times. It is a risk response strategy for overall project risks. [Planning]

PMI®, *PMBOK® Guide*, 2017, 446
PMI® *PMP Examination Content Outline*, 2015, Planning, 6, Task 10

29. c. Business case

Each project should have a business case that is used to determine whether the project should be initiated. It lists the objectives and reasons the project should be done and helps measure success at the end of the project against these objectives. One section of the business case is an analysis of the situation. This section identifies any known risks at the time the business case is prepared. Decision makers, since resources are scarce and not all proposed projects can be pursued, can review these risks a part of the decision-making process. [Planning]

PMI®, *PMBOK® Guide*, 2017, 30–31
PMI® *PMP Examination Content Outline*, 2015, Planning, 6, Task 10

30. c. Schedule baseline and cost baseline

    The cost and schedule baselines of a project are two areas significantly affected by risk occurrences. Information on these two areas, because of their quantitative nature, provides excellent input to the Perform Quantitative Risk Analysis process to help determine overall impact and to provide guidelines on establishing and managing risks. Both baselines are the starting point where the effect of individual risks can be evaluated. [Planning]

    PMI®, *PMBOK® Guide*, 2017, 430
    PMI® *PMP Examination Content Outline,* 2015, Planning, 6, Task 10

31. c. Define how best to conduct risk management activities

    Meetings are a tool and technique in the Plan Risk Management process. Typical attendees are the project manager, selected team members, key stakeholders, anyone in the organization with authority to manage the risk management planning and execution activities [such as a member of a PMO with expertise in this area or a member of the core team with this expertise], and others as needed. Customers, sellers, and regulators external to the organization may be invited. In these meetings, the risk management plan may be developed, and plans for conducting risk management activities are defined, which later are part of the risk management plan. The risk management plan may be prepared in a kick-off meeting or in a separate meeting. [Planning]

    PMI®, *PMBOK® Guide*, 2017, 404
    PMI® *PMP Examination Content Outline,* 2015, Planning, 5, Task 10

32. b. Duration estimates

    While a number of plans and project documents should be reviewed in the Identify Risk process, of the possible answers, only duration estimates are listed. They should be reviewed since they are quantitative assessments of project durations. While they often are expressed as a range, they can indicate the degree of risk associated with these estimates. A structured review of the duration estimates may indicate the current estimates could be flawed and may pose a risk to the project. [Planning]

    PMI®, *PMBOK® Guide*, 2017, 412
    PMI® *PMP Examination Content Outline,* 2015, Planning, 6, Task 10

33. b. Considers the effectiveness of the risk management process

   The purpose of the risk audit is the answer to this question. The project manager has the responsibility to see that they are performed frequently or as described in the risk management plan. They may be done during a project team meeting, as part of a risk review meeting, or as a separate risk audit meeting. Before a risk audit is conducted, the project manager should determine its format and clarify its objectives. [Monitoring and Controlling]

   PMI®, *PMBOK® Guide*, 2017, 456
   PMI® *PMP Examination Content Outline*, 2015, Monitoring and Controlling, 9, Task 4

34. a. Data quality assessment

   Perform Qualitative Risk Analysis requires accurate and unbiased data. The use of low-quality data may result in a qualitative risk analysis that is of little use to the project manager regarding understanding of the risk, data available about the risk, data quality, and data reliability and integrity. Often, the collection of risk information is difficult and consumes more time and resources than planned. It may be done through a questionnaire that measures stakeholder perceptions of various characteristics, such as completeness, objectivity, relevancy, and timeliness. Then, a weighted average of data quality characteristics can be prepared for an overall score. A risk data quality assessment is one of the data analysis tools and techniques in this process. [Planning]

   PMI®, *PMBOK® Guide*, 2017, 423
   PMI® *PMP Examination Content Outline*, 2015, Planning, 6, Task 10

35. b. Exploit

   Although it might have a negative connotation, exploitation is a strategy used for risks with positive impacts where the organization wants to ensure that the opportunity is realized. This strategy seeks to eliminate uncertainty with a particular upside risk by ensuring the opportunity happens. Other examples, in addition to that in the question, are using new technology or upgrading technology to reduce the cost and duration required to realize the project's objectives. [Planning]

   PMI®, *PMBOK® Guide*, 2017, 444
   PMI® *PMP Examination Content Outline*, 2015, Planning, 6, Task 10

36. d. Use decision making

Decision-making techniques are often used to select a risk response strategy. They also can help prioritize strategies to consider. Multi-criteria decision analysis can be used to provide a systematic approach for establishing criteria, evaluating and ranking alternatives, and making a selection. Some criteria to consider are cost, the effectiveness of the response in changing conditions, resources, time constraints, the level of impact if the risk occurs, and possible secondary risks. [Planning]

PMI®, *PMBOK® Guide*, 2017, 446
PMI® *PMP Examination Content Outline,* 2015, Planning, 6, Task 10

37. a. Conduct an alternatives analysis

Another tool and technique in Plan Risk Responses is data analysis. These techniques can help to select a preferred risk response strategy. One is to conduct an alternatives analysis. This analysis is a simple comparison of the requirements and characteristics of alternative risk response actions. It is used to make a decision about which strategy is the most appropriate to use. [Planning]

PMI®, *PMBOK® Guide*, 2017, 446
PMI® *PMP Examination Content Outline,* 2015, Planning, 6, Task 10

38. c. A risk that impacts strategic goals

The situation in this question shows time to market is essential and must be shortened; otherwise the risk will have a negative impact on the company's strategic goals. Therefore, the strategic impact of the risk is high. Even though fast tracking can result in risks, the approach also can be monitored for effective results. In the Perform Qualitative Risk Analysis process, assessment of other risk parameters is a data analysis tool and technique in which other characteristics in addition to probability and impact are considered. [Planning]

PMI®, *PMBOK® Guide*, 2017, 424
PMI® *PMP Examination Content Outline,* 2015, Planning, 6, Task 10

39. a. Use interviews

Data gathering techniques are a tool and technique in Plan Risk Responses. Within it, interviews are suggested. By using interviews, developing responses to individual and overall project risks can be done. The interview can be structured or semi-structured. In using interviews, the interviewer needs to be unbiased and promote an atmosphere of trust with the interviewee. Confidentiality should be pointed out before the interview begins so the interviewee then provides honest and unbiased suggestions for decisions. Key stakeholders and members of the project team, including the sponsor and perhaps customer representatives, may be interviewed. [Planning]

PMI®, *PMBOK® Guide*, 2017, 442
PMI® *PMP Examination Content Outline*, 2015, Planning, 5, Task 10

40. d. Avoidance

Risk avoidance involves eliminating the threat entirely or protecting it from its impact. It may involve changing some parts of the project management plan or the entire plan. The project manager may isolate the project's objectives from the risk's impact or change the objective in jeopardy, reducing the probability of its occurrence to zero. [Planning]

PMI®, *PMBOK® Guide*, 2017, 443
PMI® *PMP Examination Content Outline*, 2015, Planning, 6, Task 10

Project Procurement Management

Practice Questions

INSTRUCTIONS: Note the most suitable answer for each multiple-choice question in the appropriate space on the answer sheet.

1. Assume you are working on a complex project. In fact, it is the most complex project in your organization, so the CEO is interested in it and asks questions about its progress in staff meetings. On this project, you are outsourcing a lot of this work. This means you should—

   a. Award firm-fixed projects
   b. Hold weekly meetings with each contractor to assess progress personally
   c. Manage these multiple contracts in sequence
   d. Have a full-time contract specialist and set up a projectized structure

2. You have decided to contract for services as resources are scarce in your company. These services may range from junior-level staff to senior-level staff to people with specialized expertise to those who are generalists. Another term for contracting for service is a—

   a. T&M
   b. RTM
   c. TOR
   d. CPAF

3. Contract type selection is dependent on the degree of risk or uncertainty facing the project manager. From the perspective of the buyer, the preferred contract type in a low-risk situation is—

    a. Firm-fixed-price
    b. Fixed-price-incentive
    c. Cost-plus-fixed fee
    d. Cost-plus-incentive fee

4. The buyer has negotiated a cost-plus-incentive fee contract with the seller. The contract has a target cost of $300,000, a target fee of $40,000, a share ratio of 80/20, a maximum fee of $60,000, and a minimum fee of $10,000. If the seller has actual costs of $380,000, how much fee will the buyer pay?

    a. $104,000
    b. $56,000
    c. $30,000
    d. $24,000

5. Assume you are managing a project and have decided to outsource about 40% of the work. You expect to award three contracts for certain work packages in the WBS. The benefit of this Conduct Procurements process is that it—

    a. Determines the type of contract to award
    b. Selects a qualified seller and implements a legal agreement for delivery
    c. Enables a detailed review of the make-or-buy decisions
    d. Avoids the need for additional change requests

6. You are monitoring and controlling the three contracts on your project. You find that two of the sellers' performance is above average, but the third seller is failing to deliver as planned, and when this seller does deliver, the costs are higher than planned. You decide this seller's performance is not going to improve. Your best course of action is to—

    a. Meet with your procurement department to see if there is a seller on the qualified seller list that can provide similar services
    b. Meet with the other two sellers separately to see if one of them can do this work
    c. Issue a change request
    d. Terminate the contract

7. You issued a RFP, and when the proposals were received, there were expected differences in the price and managerial approach sections. What was surprising was the range of different technical approaches that were offered. You then decided to—

a. Perform a review of the SOW
b. Assess if the sellers misunderstood the SOW
c. Evaluate the proposals
d. Determine if the wrong contract type was chosen

8. You are a contractor for a state agency. Your company bid on a high-value project with a firm-fixed price contract for a water resource management project for your state. Before the agency selected a seller, its project manager decided to—

a. Ensure there is senior management approval before the contract award
b. Use an expert with negotiation skills
c. Escalate the decision to your procurement department
d. Add in the terms and conditions a clause to use arbitration before issuing a change order

9. You are working to prepare your procurement management plan. You are planning to determine the type of contract to issue, but first you want to see possible costs as your University is updating its software used by students and faculty. You decide to—

a. Obtain legal advice
b. Assess marketplace conditions
c. Advertise the opportunity
d. Determine the bid type to use

10. You are awarding a contract to help develop a new feature in the 12th generation phone. You feel this feature will enable your company to outpace the competition. Since you want to outdistance the competition, you want the contractor to have the best and brightest people on its staff. As you select source selection criteria, you should include—

a. Specific relevant experience
b. Overall life-cycle costs
c. Intellectual property rights
d. Warranties

11. You are working to determine the type of contact for your data warehousing project and you want to have a number of sellers bid on this opportunity However, the project is not located in a desirable area and is it is in a remote part of your country. You decide to gauge interest and decide to—

    a. Use advertising in trade journals
    b. Set measurable performance indicators
    c. Award a contract to a systems integration contractor and expand its scope
    d. Issue an Request for Information

12. A number of administrative activities are important in procurement management. For example, you should retain—

    a. Assumptions statements
    b. Measurable procurement performance indicators
    c. Facilitated negotiations
    d. Settled subcontracts

13. Recent data indicate that more than 10,000 airline passengers are injured each year from baggage that falls from overhead bins. You performed a make-or-buy analysis and decided to outsource an improved bin design and manufacture. The project team needs to develop a list of qualified sources. As a general rule, which method would the project team find especially helpful—

    a. Advertising
    b. Internet
    c. Trade catalogs
    d. Relevant local associations

14. You are working to monitor and control your three contacts on your program. Your company is certified by the Software Engineering Institute's Capability Maturity Model for Integration, which means your three contactors also hold this certification. Of the following tools and techniques to use, you select—

    a. Contract document for the contract being closed
    b. Procurement audits reports
    c. Earned value
    d. Contract changes

15. You are working on a new project in your organization. You need to decide how best to staff the project and handle all its resource requirements. You decide to—

   a. Conduct a make-or-buy analysis
   b. Conduct a market survey
   c. Solicit proposals from sellers using an RFP to determine whether you should outsource the project
   d. Review your procurement department's qualified-seller lists and send an RFP to selected sellers

16. Your company decided to award a contract for project management services on a pharmaceutical research project. Because your company is new to project management and does not understand the full scope of services that may be needed under the contract, it is most appropriate to award a—

   a. Firm-fixed-price contract
   b. Fixed-price-incentive contract
   c. Cost-plus-a-percentage-of-cost contract
   d. Time-and-materials contract

17. Requirements for warranty and future product support are found in the—

   a. Proposal
   b. Statement of work
   c. Contract terms and conditions
   d. Agreement

18. You plan to award a contract to provide project management training for your company. You decide it is important that any prospective contractor have an association with a major University that awards master's certificates in project management. This is an example of—

   a. Setting up an independent evaluation
   b. Preparing requirements for your statement of work
   c. Establishing a weighting system
   d. Establishing source selection criteria

19. The project team has the responsibility to ensure a procurement agreement—

   a. Incorporates other items the seller specifies
   b. Focuses on all selected sellers deemed to be in the competitive range
   c. Adheres to the negotiation process
   d. Adheres to the organization's procurement policies

20. During the course of working with a seller that no one had worked with previously and is working on site with your team, in preparing the contract you specified that inspections and audits can be conducted. You decided to hire an outside person, considered to be an expert in project management to conduct an audit. In doing so, it is important to ensure that—

   a. The audit can include the company's management personnel
   b. The audit should be completely quickly, so you can process change requests as needed for recommendations
   c. The auditor focuses on completed work, not work in progress
   d. The auditor has credibility

21. Assume you have completed your make-or-buy analysis and decided to outsource most of the development of safe flying cars to ease traffic congestion in your city. You are determining your procurement strategy. One of its objectives is to—

   a. Determine the type of legally binding agreement
   b. Establish the appropriate contract type
   c. Decide if a bidders conference should be held
   d. Develop a statement of work

22. Specifications, quality levels, desired quality, performance data, performance period, and location are items to—

   a. Discuss with sellers in contract negotiations
   b. Stated in a Request for Proposal
   c. Include in a procurement statement of work
   d. Include in the final agreement

23. An often-overlooked output of the Conduct Procurements process is—

   a. Analytical techniques
   b. Resource calendars
   c. Work performance information
   d. Work performance data

24. Assume you have completed your procurement management plan and your make-or-buy analysis and now are updating project documents as a result of this work. You need to update the—

   a. Stakeholder analysis
   b. Requirements traceability matrix
   c. Negotiations process
   d. Management systems for contractual relationships

25. You have decided to award a contract to a seller that has provided quality services to your company frequently in the past. Your current project, although somewhat different from previous projects, is similar to other work the seller has performed. In this situation, to minimize your risk you should award what type of contract?

    a. Fixed price with economic price adjustment
    b. Fixed-price-incentive
    c. Firm-fixed-price
    d. Cost-plus-award-fee

26. As project manager, you need a relatively fast and informal method addressing disagreements with contractors. One such method is to submit the issue in question to an impartial third party for resolution. This process is known as—

    a. Alternative dispute resolution
    b. Problem processing
    c. Steering resolution
    d. Mediation litigation

27. Work performance data often are used in Control Procurements to show—

    a. Whether quality standards are being met
    b. If approved change requests are being implemented
    c. Current or potential problems
    d. If seller invoices have been paid

28. In the Plan Procurement Management process, potential sellers are evaluated especially if the buyer wants to exercise some degree of influence or control over acquisition decisions. This means thought also should be given to—

    a. Cost estimates developed by potential sellers
    b. The scope baseline
    c. How well sellers implement corrective actions as needed
    d. Responsibility to acquire professional licenses

29. A buyer has negotiated a fixed-price-incentive-fee contract with the seller. The contract has a target cost of $200,000, a target profit of $30,000, and a target price of $230,000. The buyer also has negotiated a ceiling price of $270,000 and a share ratio of 70/30. If the seller completes the contract with actual costs of $170,000, how much profit will the buyer pay the seller?

    a. $21,000
    b. $35,000
    c. $39,000
    d. $51,000

30. Requirements for acceptance criteria are defined in the—

    a. Agreement
    b. Procurement management plan
    c. Overall project management plan
    d. Specifications

31. Sellers of course want to be paid once they have delivered what they promised in the agreement. This means payment information is—

    a. Managed by contracting officer's technical representative responsibility
    b. Stated in the procurement documentation
    c. Performed by the accounts payable system
    d. Managed by the procurement department

32. Assume before you award a contract, you and your team are striving to identify areas that may have more risk. You decide to—

    a. Use an independent estimator
    b. Evaluate past performance
    c. Review the risk management plan
    d. Review the source selection criteria

33. Assume in your project, you need to use a contractor. But you and the seller have a prior agreement in place. This means together you can prepare a procurement statement of work. It is an example of—

    a. Core capabilities of the organization
    b. Organizational process assets
    c. Mutual development of contract clauses
    d. Value delivered by vendors

34. You work for an aerospace and defense contractor, and you plan to outsource 75% of the work as you develop the next generation passenger jet. Because so much outsourcing will be used, you decide to use a systems integration contractor who will integrate and oversee the work of the other contractors. In this situation, you plan to use a—

    a. Fixed-price with economic price adjustment contract
    b. Cost-plus-award-fee contract
    c. Cost-price-incentive-fee-contract
    d. Sole source agreement

35. Although you finished your procurement management plan, you and your team decided since the project is so complex that you would be outsourcing about 50% of the work to be done on your new drug to cure macular degeneration. Given the large procurement, you decided it would be helpful to—

    a. Use partnering with the selected sellers
    b. Have an independent estimate
    c. Establish a procurement change control system
    d. Structure each resulting agreement to minimize claims

36. You are working on a complex project designed to combat glaucoma without the need to take eye drops. You are going to outsource part of the work to experts in the glaucoma field and have identified four sellers that are capable and respected. These sellers—

    a. Have negotiated a draft contract
    b. Will give presentations to your executive team
    c. Will established partnering agreements to reduce the possibility of claims
    d. Are in the competitive range

37. Assume you thought you would need to outsource part of your work on your telecom project. You conducted a make-or-buy analysis, and the decision was made not to outsource, This means—

    a. You have no additional work remaining in procurement management
    b. The decision should be documented in a decision log
    c. You should update organizational process assets
    d. You should review the make-or-buy analysis you conducted

38. You are the project manager with a team to design the next generation of automobiles for your company. After an extensive make-or-buy analysis, you have determined which items will be outsourced. Now you are working on the statements of work for each item to be procured. It is important as you do so to—

    a. Determine constraints and assumptions
    b. Follow the scope management plan
    c. Identify any prequalified sellers
    d. Recognize long lead items

39. Which of the following types of contracts has the least risk to the seller?

    a. Firm-fixed-price
    b. Cost-plus-fixed-fee
    c. Cost-plus-award-fee
    d. Fixed-price-incentive-fee

40. Assume that your company has a cost-plus-fixed-fee contract. The contract value is $110,000, which consists of $100,000 of estimated costs with a 10-percent fixed fee. Assume that your company completes the work but only incurs $80,000 in actual cost. What is the total cost to the project?

    a. $80,000
    b. $90,000
    c. $10,0000
    d. $125,000

Answer Sheet

1.	a	b	c	d		21.	a	b	c	d
2.	a	b	c	d		22.	a	b	c	d
3.	a	b	c	d		23.	a	b	c	d
4.	a	b	c	d		24.	a	b	c	d
5.	a	b	c	d		25.	a	b	c	d
6.	a	b	c	d		26.	a	b	c	d
7.	a	b	c	d		27.	a	b	c	d
8.	a	b	c	d		28.	a	b	c	d
9.	a	b	c	d		29.	a	b	c	d
10.	a	b	c	d		30.	a	b	c	d
11.	a	b	c	d		31.	a	b	c	d
12.	a	b	c	d		32.	a	b	c	d
13.	a	b	c	d		33.	a	b	c	d
14.	a	b	c	d		34.	a	b	c	d
15.	a	b	c	d		35.	a	b	c	d
16.	a	b	c	d		36.	a	b	c	d
17.	a	b	c	d		37.	a	b	c	d
18.	a	b	c	d		38.	a	b	c	d
19.	a	b	c	d		39.	a	b	c	d
20.	a	b	c	d		40.	a	b	c	d

Answer Key

1. c. Manage these multiple contracts in sequence

   Complex projects often involve contractors and subcontractors. While they may be managed in sequence, they also may be managed simultaneously. In such cases, each contract life cycle may begin and end during any phase of the project life cycle. They buyer-seller relationship may exist at many levels on any project, and between organizations internal or external to the acquiring organization. [Monitoring and Controlling]

   PMI®, *PMBOK® Guide*, 2017, 401
   PMI® *PMP Examination Content Outline*, 2015, Monitoring and Controlling, 9, Task 7

2. c. TOR

   The TOR means terms of reference and is another term that can be used to contract for services. It is similar to a procurement statement of work and describes the tasks the contractor will perform, standards the contractor will fulfill, data to submit for approval, services the contractor will provide, and the schedule for the initial submission and the review and approval time needed. [Planning]

   PMI®, *PMBOK® Guide*, 2017, 478
   PMI® *PMP Examination Content Outline*, 2015, Planning, 6, Task 7

3. a. Firm-fixed-price

   Buyers prefer the firm-fixed-price contract because it places more risk on the seller. Although the seller bears the greatest degree of risk, it also has the maximum potential for profit. Because the seller receives an agreed-upon amount regardless of its costs, it is motivated to decrease costs by efficient production. It is the most commonly used contract as the price for goods is set at the beginning and is not changed unless the scope of work changes. Types of contracts are an input to the Plan Procurement Management process. [Planning]

   PMI®, *PMBOK® Guide*, 2017, 471
   PMI® *PMP Examination Content Outline*, 2015, Planning, 6, Task 7

4. d. $24,000

Comparing actual costs with the target cost shows an $80,000 overrun. The overrun is shared 80/20 (with the buyer's share always listed first). In this case 20% of $80,000 is $16,000, the seller's share, which is deducted from the $40,000 target fee. The remaining $24,000 is the fee paid to the seller. In this type of contract, the seller is reimbursed for all allowable costs for performing the contract work and receives a predetermined incentive fee based on achieving certain performance objectives set forth in the contract. If the final costs are less than or greater than the original estimated costs, then the buyer and seller share costs based on a predetermined cost sharing formula, such as the 80/20 split over/under target costs based on the seller's actual performance. [Planning]

PMBOK® Guide, 2017, 472
PMI® *PMP Examination Content Outline*, 2015, Planning, 6, Task 7

5. b. Selects a qualified seller and implements a legal agreement for delivery

The Conduct Procurements process purpose is to obtain seller responses, select a seller, and award contracts. During the process, the team receives bids or proposals and will apply previously defined selection criteria to select the sellers qualified to perform the work. The benefit of this process is the answer to this question.

PMBOK® Guide, 2017, 482
PMI® *PMP Examination Content Outline*, 2015, Executing, 7, Task 2

6. c. Issue a change request

It is obvious this contract will need to be terminated, and termination clauses are in the procurement management plan. However, before terminating the contract, you will need to issue a change request, an output of this process. Requested but unresolved changes by the buyer or actions by the seller may be disputed by one party, meaning these changes are identified and documented by project correspondence and other documentation. [Monitoring and Controlling]

PMBOK® Guide, 2017, 499
PMI® *PMP Examination Content Outline*, 2015, Monitoring and Controlling, 9, Task 7

7. c. Evaluate the proposals

A data analysis technique in Conduct Procurement is proposal evaluation. They are evaluated to ensure they are complete, respond to the bid documents, the procurement statement of work, source selection criteria, and other documents that may be in the bid documents. [Executing]

PMI®, *PMBOK® Guide,* 2017, 487
PMI® *PMP Examination Content Outline,* 2015, Executing, 8, Task 2

8. a. Ensure there is senior management approval before the contract award

Selected sellers are an output of Conduct Procurements. These are sellers that have been judged to be in the competitive range. However, final approval of a high-value or high-risk procurement as in this question tends to require senior management approval before contract or agreement award. [Executing]

PMI®, *PMBOK® Guide,* 2017, 488
PMI® *PMP Examination Content Outline,* 2015, Executing, 9, Task 2

9. b. Assess marketplace conditions

Marketplace conditions and legal advice are both enterprise environmental factors to consider in preparing the procurement management plan and in determining the type contract to award. Marketplace conditions can show cost of similar updates at other Universities near you or in your same region. Another enterprise environmental factor that is appropriate is to see products and services available in the marketplace. [Planning]

PMI®, *PMBOK® Guide,* 2017, 470
PMI® *PMP Examination Content Outline,* 2015, Planning, 6, Task 7

10. a. Specific relevant experience

An output of Plan Procurement Management is source selection criteria. There are a variety of criteria to consider. Specific relevant experience is a criterion to include. Also, technical expertise and approach and the staff's qualifications, availability, and competence should be considered. [Planning]

PMI®, *PMBOK® Guide,* 2017, 478
PMI® *PMP Examination Content Outline,* 2015, Planning, 6, Task 7

11. d. Request for Information

Procurement or bid documents are used to solicit proposals from prospective sellers. A Request for Information is used when more information on the goods and services to be acquired is needed from the sellers. It then is followed by a Request for Quotation or a Request for Proposal. It is an output of Plan Procurement Management. [Planning]

PMI®, *PMBOK® Guide,* 2017, 477
PMI® *PMP Examination Content Outline,* 2015, Planning, 6, Task 7

12. b. Measurable procurement performance indicators

Control Procurements includes application of project management processes to contractual relationships. Within this process are several administrative activities. One is to collect data and manage project records. It includes maintaining detailed records on physical financial performance and establishing measurable procurement indicators. [Monitoring and Controlling]

PMI®, *PMBOK® Guide,* 2017, 494
PMI® *PMP Examination Content Outline,* 2015, Monitoring and Controlling, 9, Task 7

13. a. Advertising

Advertising is a tool and technique in Conduct Procurements. It enables existing lists of sellers to be expanded by using advertising in general circulation such as in newspapers or specialty trade publications. Online resources also can be used. Some government agencies require public advertising for certain types of procurement items, and most require public advertising or online posting of pending government contracts. [Executing]

PMI®, *PMBOK® Guide,* 2017, 476
PMI® *PMP Examination Content Outline,* 2015, Executing, 8, Task 2

14. c. Earned value

Earned value analysis is a data analysis tool and technique in Control Procurement. In this situation, since the contractor and subcontractors are Level 3 certified for this agreement, they are using earned value. In Control Procurement it provides schedule and cost variances along with calculations of schedule and cost indices to determine the degree of variance from the target. [Monitoring and Controlling]

PMI®, *PMBOK® Guide,* 2017, 498
PMI® *PMP Examination Content Outline,* 2015, Monitoring and Controlling, 9, Task 7

15. a. Conduct a make-or-buy analysis

A make-or-buy analysis is a Plan Procurement Management tool and technique in data analysis used to determine whether a product, service, or result can be produced or performed cost effectively by the performing organization or should be contracted out to another organization. Factors to consider include the organization's resource allocation and their skills and abilities, the need for special expertise, the desire not to expand permanent employment obligations, and the need for individual expertise. [Planning]

PMI®, *PMBOK® Guide,* 2017, 473
PMI® *PMP Examination Content Outline,* 2015, Planning, 6, Task 7

16. d. Time-and-materials contract

A time-and-materials contract is a type of contract that provides for the acquisition of supplies or services on the basis of direct labor hours, They are a hybrid type of contract in that they contain aspects of cost-reimbursable and fixed-price contracts. They are often used to augment existing staff, acquire experts, and obtain outside support when a precise statement of work cannot be quickly prepared. They fit the scenario in this question and can increase in contract value if required. They also are part of organizational process assets, an input to Plan Procurement Management. [Planning]

PMI®, *PMBOK® Guide,* 2017, 472
PMI® *PMP Examination Content Outline,* 2015, Planning, 6, Task 7

17. d. Agreement

Agreements are an output of Conduct Procurements. They are mutually binding agreements and obligate the seller to provide specified products, services, or results. The buyer compensates the seller, and the agreement then is a legal relationship subject to remedy in the court. While each agreement is different and contains different items, some provide warranties and future product support. For example, if the agreement is for software services, often warranties and future product support are included. [Executing]

PMI®, *PMBOK® Guide,* 2017, 489
PMI® *PMP Examination Content Outline,* 2015, Executing, 8, Task 1

18. d. Establishing source selection criteria

   The selection criteria are typically included in procurement documents and are then used to rate or score seller responses to ensure the proposal that is selected will offer the best quality for the services to be provided. In this specific question, they are more than just price as they also include other technical and management areas. The criteria may be set up as a numerical score, a color code, or a written description of the seller's understanding of the buyer's needs. They may be used to select a single seller or to establish a negotiating sequence. Source selection criteria are an output of Plan Procurement Management. [Planning]

   PMI®, *PMBOK® Guide*, 2017, 478–479
   PMI® *PMP Examination Content Outline*, 2015, Planning, 6, Task 7

19. d. Adheres to the organization's procurement policies

   Agreements are an output of the Conduct Procurements process. However, an input in this process is organizational process assets. One of these organizational process assets is adherence to organizational policies that influence seller selection. These policies need to be reviewed before an agreement. [Executing]

   PMI®, *PMBOK® Guide*, 2017, 486
   PMI® *PMP Examination Content Outline*, 2015, Executing, 8, Task 2

20. d. The auditor has credibility

   In Control Procurements the quality of the controls including the independence and credibility of procurement audits are critical to success. Controls using techniques such as these audits need to be viewed in a positive way. Since the auditors are independent, they are unbiased, and their findings and recommendations can improve the overall success of procurement management. They are structured reviews of the procurement process, and their observations should be presented to both the buyer's and seller's project managers for any needed adjustments. [Monitoring and Controlling]

   PMI®, *PMBOK® Guide*, 2017, 494, 498
   PMI® *PMP Examination Content Outline*, 2015, Monitoring and Controlling, 9, Task 7

21. a. Determine the type of legally binding agreement

   There are several objectives of the procurement strategy. One is the answer to the question. Other objectives are to determine the project delivery method, and another is to determine how the procurement will advance throughout the procurement system. [Planning]

   PMI®, *PMBOK® Guide,* 2017, 478
   PMI® *PMP Examination Content Outline,* 2015, Planning, 6, Task 7

22. c. Include in a procurement statement of work

   The procurement statement of work is an output of Plan Procurement Management. It is developed from the scope baseline and defines the portion of the scope of the project to be included in related agreements. It describes the procurement items in detail, so the sellers can determine if they can provide the needed products, services, and goods. Examples of items that can be included are the ones in this question. [Planning]

   PMI®, *PMBOK® Guide,* 2017, 477
   PMI® *PMP Examination Content Outline,* 2015, Planning, 6, Task 7

23. b. Resource calendars

   They are an output of Conduct Procurements as the quantity and availability of contracted resources and the dates on which each specific resource or resource group can be active or is idle is documented. They are part of the project documents to update. [Executing]

   PMI®, *PMBOK® Guide,* 2017, 491
   PMI® *PMP Examination Content Outline,* 2015, Executing, 8, Task 2

24. b. Requirements traceability matrix

   Other documents to update are the requirements documents, lessons learned repository, stakeholder register, milestone list, and the risk register. The procurement management plan will include certain requirements that may not have been considered at the time the traceability matrix was prepared. It is important in that it links product requirements from their origin to the deliverables that satisfy them. Each procurement statement of work, for example, contains requirements that the seller must meet; other requirements, such as performance bonds or insurance contracts and direction to sellers on developing a WBS are part of the procurement management plan. [Planning]

   PMI®, *PMBOK® Guide,* 2017, 480
   PMI® *PMP Examination Content Outline,* 2015, Planning, 6, Task 7

25. c. Firm-fixed-price

In a firm-fixed-price contract, the seller receives a fixed sum of money for the work performed regardless of costs. This arrangement places the greatest financial risk on the seller and encourages it to control costs. Any cost increase because of adverse performance is the seller's responsibility, who is obligated to complete the effort. The buyer should precisely state the products or services to be procured as any changes to the procurement specification can increase the buyer's cost. It is the best choice in this question since this seller is entering into a new area of work. [Planning]

PMI®, *PMBOK® Guide,* 2017, 471
PMI® *PMP Examination Content Outline,* 2015, Planning, 6, Task 7

26. a. Alternative dispute resolution

Alternative dispute resolution is used in Control Procurements in contested changes. If the buyer and seller cannot reach an agreement on contested changes or that a change has occurred, claims result. It then may be handled through alternative dispute resolutions. It is a relatively informal way to address differences of opinion. Its purpose is to address such issues without having to seek formal legal redress through the courts. [Monitoring and Controlling]

PMI®, *PMBOK® Guide,* 2017, 498
PMI® *PMP Examination Content Outline,* 2015, Monitoring and Controlling, 9, Task 7

27. d. If seller invoices have been paid

Work performance data are an input to Control Procurements. Additionally, these data show costs that have been incurred or committed, and technical performance activities that have started, are in progress, or have completed. [Monitoring and Controlling]

PMI®, *PMBOK® Guide,* 2017, 496
PMI® *PMP Examination Content Outline,* 2015, Monitoring and Controlling, 9, Task 7

28. d. Responsibility to acquire professional licenses

An input to Conduct Procurement is requirements documentation. It discusses requirements with contractual and legal implications. They may include health, security, performance, environmental, insurance, intellectual property rights, licenses, permits, and any other non-technical requirements. [Executing]

PMI *PMBOK® Guide,* 2017, 485
PMI® *PMP Examination Content Outline,* 2015, Executing, 8, Task 2

29. c. $39,000

To calculate the fee that the buyer must pay, actual costs are compared with the target cost. If actual costs are less than the target cost, the seller will earn profit that is additional to the target profit. If actual costs are more than the target cost, the seller will lose profit from the target profit. The amount of profit is determined by the share ratio (with the buyer's share listed first). In this example, the seller is under target cost by $30,000. That amount will be split 70/30. So the buyer keeps $21,000, and the seller receives an additional $9,000 added to the target profit, which is the incentive. Total fee is $39,000. In this type of contract the buyer and seller have flexibility in that it allows for deviations from performance with financial incentives in place tied to agreed-upon metrics. These incentives tend to be related to cost, schedule, or technical performance. The performance targets are established at the beginning. The price ceiling is set, and all costs above the price are the seller's responsibility. [Planning]

PMI®, *PMBOK® Guide,* 2017, 471
PMI® *PMP Examination Content Outline,* 2015, Planning, 6, Task 7

30. a. Agreement

Agreements are an output of Conduct Procurements. They are a mutually binding agreement in which the seller provides the products, services, or results, and the buyer compensates the seller. There are many components of an agreement, one is acceptance criteria. It is described along with quality and inspection. [Executing]

PMI®, *PMBOK® Guide,* 2017, 489
PMI® *PMP Examination Content Outline,* 2015, Executing, 8, Task 2

31. b. Stated in the procurement documentation

Procurement documentation is a tool and technique in Control Procurements. They contain complete supporting records to administer the procurement processes. They include the statement of work, payment information, contractor work performance information, plans, drawings, and correspondence. [Monitoring and Controlling]

PMI®, *PMBOK® Guide,* 2017, 496
PMI® *PMP Examination Content Outline,* 2015, Monitoring and Controlling, 9, Task 7

32. c. Review the risk management plan

The risk management plan is one of several components of the project management plan, an input to Conduct Procurement. The risk management plan should be reviewed since it describes how risk management is conducted in the project. For example, the type of agreement that is selected has its own degree of risk to the seller and to the buyer. Other parts of the agreement further can be more stringent based on the level of risk threshold or more lenient, especially if the buyer has worked with the seller successfully on other projects. [Executing]

PMI®, *PMBOK® Guide*, 2017, 484
PMI® *PMP Examination Content Outline*, 2015, Executing, 6, Task 2

33. b. Organizational process assets

They are an input to Plan Procurement Management. This example is one of a preapproved seller list. They are lists of sellers that have been previously vetted by the buyer. They are used to streamline the procurement process and enable the buyer and seller to work closely together. They can shorten the timeline for the seller selection process. [Planning]

PMI®, *PMBOK® Guide*, 2017, 471
PMI® *PMP Examination Content Outline*, 2015, Planning, 7, Task 7

34. b. Cost-plus-award-fee contract

While there are a variety contract types to use, this situation is a perfect example of the cost-plus-award-fee contract. In it the seller is reimbursed for legitimate costs. The majority of the fee is earned based on the buyer satisfaction with broad, subjective performance criteria. These criteria are defined and incorporated into the contract. The determination of the fee is based on buyer's subjective assessment of the seller's performance and typically cannot be appealed. [Planning]

PMI®, *PMBOK® Guide*, 2017, 472
PMI® *PMP Examination Content Outline*, 2015, Planning, 6, Task 7

35. b. Have an independent estimate

    On large procurements the buyer may decide to prepare its own independent estimate or may have an outside professional prepare an estimate. The purpose is to have a benchmark for proposed responses. If there are significant differences in cost estimates, it may be an indication that the procurement statement of work is deficient or ambiguous or that the prospective sellers have misunderstood or failed to respond fully to the procurement statement of work. [Planning]

    PMI®, *PMBOK® Guide*, 2017, 479
    PMI® *PMP Examination Content Outline*, 2015, Planning, 6, Task 7

36. d. Are in the competitive range

    Selected sellers are an output of Conduct Procurements. These selected sellers are ones in a competitive range based on the proposal or bid evaluation. Typically approval of high-value, high-risk, complex procurements generally will require approval by senior managers prior to award. [Executing]

    PMI®, *PMBOK® Guide*, 2017, 488
    PMI® *PMP Examination Content Outline*, 2015, Executing, 8, Task 2

37. d. You should review the make-or-buy analysis you conducted

    Make-or-buy decisions are an output of Plan Procurement Management. As in this question, the team may decide to do the work, or the needs may be purchased from external sources. [Planning]

    PMI®, *PMBOK® Guide*, 2017, 479
    PMI® *PMP Examination Content Outline*, 2015, Planning, 5, Task 7

38. b. Follow the scope management plan

    The procurement statement of work is an output of Plan Procurement Management. The statement of work for each procurement uses the scope management plan. It is an input to this process under the project management plan, It describes how the scope of work will be managed though the execution of the project. The other possible answers are items that are part of the procurement management plan. [Planning]

    PMI®, *PMBOK® Guide*, 2017, 409
    PMI® *PMP Examination Content Outline*, 2015, Planning, 6, Task 7

39. b. Cost-plus-fixed-fee

On a firm-fixed-price contract, the seller absorbs 100 percent of the risks; while on a cost-type contract, the buyer carries the most risk. Cost-reimbursable contracts provide flexibility to the project to redirect the seller whenever the scope of work cannot be precisely defined at the start and needs to be altered or when high risks may exist in the effort. Cost-plus-fixed-fee contracts have less risk to sellers than cost-plus-award-fee or cost-plus-incentive-fee contracts because the fee is fixed based on costs, so the seller is guaranteed a certain level of profit. The fee is paid only for completed work and does not change unless the project scope changes. [Planning]

PMI®, *PMBOK® Guide*, 2017, 472
PMI® *PMP Examination Content Outline*, 2015, Planning, 6, Task 7

40. b. $90,000

In this situation the fixed-fee of $10,000 does not change but now represents a seller profit of 12.5 percent on incurred costs. This means that the total cost to the project is $90,000. The actual fee is determined when the seller completes its work. [Planning]

PMI®, *PMBOK® Guide*, 2017, 472
PMI® *PMP Examination Content Outline*, 2015, Planning, 6, Task 7

Project Stakeholder Management

Practice Questions

INSTRUCTIONS: Note the most suitable answer for each multiple-choice question in the appropriate space on the answer sheet.

1. During your project, you will have a number of different types of meetings. Some will be informational, others will be key updates, and some will be for decision-making purposes. While different attendees will attend each meeting, a best practice to follow is to:

   a. Group stakeholders into categories to determine which ones should attend each meeting
   b. Invite those stakeholders who have a high level of interest in your project to attend each meeting
   c. Be sensitive to the fact that stakeholders often have very different objectives
   d. Recognize that roles and responsibilities may overlap but practice a policy of 'no surprises' and inform your stakeholders about any upcoming meetings

2. You are managing a project with team members located at customer sites on three different continents. You have a number of stakeholders on your project, and most of them are located outside of the corporate office. As you work to monitor and control Stakeholder Engagement, you should—

   a. Consult the organizational process assets
   b. Consult the enterprise environmental factors
   c. Use assumptions analysis
   d. Review the change log

3. When you are preparing your project charter, you also are identifying your stakeholders although you continue the process throughout the project. As you are doing so, you decide to do some stakeholder analysis. You decide to classify stakeholders based on directions of influence, which means you are—

   a. Comparing power and influence
   b. Using a stakeholder cube
   c. Comparing upward, downward, outward, and sideward approaches
   d. Using prioritization

4. You are responsible for a project in your organization that has multiple internal customers. Because many people in your organization are interested in this project, you decide to prepare a stakeholder engagement strategy. Before preparing this strategy, you first should—

   a. Conduct a stakeholder analysis to collect information
   b. Determine a production schedule to show when each stakeholder needs each type of information produced
   c. Determine the potential impact that each stakeholder may generate
   d. Prioritize each stakeholder's level of interest and influence

5. Recognizing the importance of preparing a stakeholder engagement plan, you met with your team to obtain their buy in and to discuss it. You explained the key benefit of Plan Stakeholder Engagement is to—

   a. Determine appropriate strategies for a continual focus on identifying stakeholders throughout the life cycle
   b. Provide a clear plan that is actionable to interact effectively with stakeholders
   c. Develop appropriate management strategies to effectively engage stakeholders
   d. Plan a series of meetings to ensure stakeholders remain interested and to address their concerns

6. Assume you are actively working, along with your team, to manage stakeholder engagement on your project to develop a new drug to prevent any retina problems of any type. You know you must manage their engagement throughout the project life cycle. This means some project documents will need updating including—

   a. Informal and formal project reports
   b. The stakeholder register
   c. The stakeholder engagement plan
   d. The communications plan

7. Stakeholders often have issues, and you have asked each of your team members to document them. At each team meeting, you and your team discuss them and determine appropriate responses. You have a project issue log, which is—

   a. Part of the project's lessons learned
   b. Added to the stakeholder register to show which stakeholder raised it
   c. An output from the Manage Stakeholder Engagement process
   d. An output from the Control Stakeholder Engagement process

8. As you work on your project to update its software training classes to focus on an agile approach, you have a number of key stakeholders. As many students and their managers are requesting these classes, your CEO has taken a special interest in your project and has asked you to accelerate your schedule to complete it in two months rather than in your planned six months but still have quality offerings. This means as you work to monitor overall project stakeholder relationships, you should—

   a. Provide notifications to stakeholders about status regularly
   b. Ask your stakeholders for regular feedback as you work on your project
   c. Provide presentations to each stakeholder group
   d. Determine how changes will be monitored and controlled

9. As you work with your team to prepare your stakeholder engagement plan, you decided to develop a stakeholder engagement assessment matrix. You set it up, so you could—

   a. Show the phase of your project of interest to identified stakeholders
   b. Show gaps between current and desired levels of engagement
   c. Determine which stakeholders you and your team felt were critical to project success but did not know about it
   d. Determine when to involve key stakeholders in your project

10. Assume you are working to prepare your stakeholder engagement plan. You want it to reflect the diverse information needs of your project stakeholders as you prepare your plan. However, you recognize you will need to update it because your stakeholders will change especially since your project will last two years. A trigger situation that shows it is time to update the plan is—

    a. When assumptions change
    b. If your project charter changes
    c. When the organization structure changes
    d. If the time frame and frequency for the distribution of required information to stakeholders changes

11. Having worked as a project manager for nine years, you know how important it is to identify the critical stakeholders, so you do not overlook anyone who has a major influence on your project even if you do not ever plan to meet with or talk with this individual. As you work with your team, you explain the key benefit of the Identify Stakeholder Process is that it—

    a. Identifies the people, groups, or organizations that could impact or influence project decisions
    b. Shows the interdependencies among project stakeholders to enable classification for how best to involve them on your project
    c. Identifies the appropriate focus for each stakeholder or a group of stakeholders
    d. Shows the potential impact each stakeholder has on project success

12. Assume you are working to Identify Stakeholders and are now gathering information about the stakeholders on your project. You decide to allow people who will participate in your brainstorming session to consider the questions you plan to ask before you meet as a group. This approach is—

    a. The nominal group technique
    b. Used to assess how stakeholders probably will respond in various situations
    c. Brain writing
    d. Mind mapping

13. You realize that on projects, some stakeholders will not be as receptive as others to your project and actually can be negative from the beginning. Assume you have classified your stakeholders on your project designed to ensure students in your city have access to the best educational resources available, whether in class or on line. Now you are preparing your stakeholder engagement plan, and you want to determine the appropriate strategy to improve each stakeholder' level of engagement. You decide to—

    a. Review the validity of the underlying constraints
    b. Use root-cause analysis
    c. Review the validity of its underlying assumptions
    d. Use mind mapping

14. Stakeholder engagement involves a number of critical activities. An example is—

    a. Managing stakeholder expectations by using negotiation and communications
    b. Developing management strategies to engage them during the project's life cycle
    c. Adjusting strategies and plans to engage stakeholders effectively
    d. Identifying the scope and impact of changes to project stakeholders

15. Work performance information is an output of Monitor Stakeholder Engagement. It includes a number of items, one of which is—

    a. Change requests
    b. Issue log
    c. Documented lessons learned
    d. Status of stakeholder engagement

16. Often in working as a project manager, it is easy to overlook key stakeholders. Assume you work for a device manufacturer and are working as the project manager for the next generation valve replacement. Your company has been a leader in this market, which means you have a lot of lessons learned available to you. Your project is scheduled to last four years. As a best practice, you should—

    a. Work actively with your company's Knowledge Management Officer
    b. Consult regularly with your program manager
    c. Work actively with members of your Governance Board
    d. Work actively with members of your company's Portfolio Review Board

17. Assume you are managing the development of a construction project in your city to replace its five bridges so they are state of the art and meet updated safety standards since they originally were constructed 20 years ago. The design work has been completed, you have awarded subcontracts, and are set to begin construction. Today your legal department told you to stop work as you had not consulted them, and there was a critical standard you overlooked during the design process. This example shows—

    a. You need to continually work to engage stakeholders on your project
    b. You should use a RACI chart and have one of your team members work with the legal department throughout the project
    c. You should provide the legal department with a copy of your stakeholder engagement plan and ask for their representative to sign it and offer any comments
    d. You need to continually identify project stakeholders

18. Assume your construction project is for a small city with only 8,500 people. There has been opposition to it from the beginning, when the City Commissioners approved it by many residents. The residents recognize they will be severely impacted as the new bridges are implemented, and during the public hearings before the Commissioners' decision, they hired an attorney to state they felt the more cost-effective approach was to strengthen the bridges, so they meet today's safety requirements. Residents now know you have been ordered to stop work, and they have requested a meeting with the Commission on Tuesday. This means you should—

    a. Develop a mitigation plan to present at this meeting
    b. Work diligently with the legal department to satisfy their concerns and receive a go ahead before Tuesday's meeting
    c. Demonstrate at the meeting the sustainability impacts of the new bridges
    d. Balance the interests of these negative stakeholders and meet with them before Tuesday's meeting

19. The salience model is one way to classify stakeholders. In it—

    a. Stakeholders' power, urgency, and legitimacy are used
    b. Stakeholders' level of authority and concern are used
    c. Stakeholders' active involvement and power are used
    d. Stakeholders' influence and ability to effect changes are used

20. In Plan Stakeholder Engagement, a number of organizational assets are used as inputs, which include—

    a. Organization culture and the political climate
    b. Practices and habits and templates
    c. Lessons learned repository
    d. Organization's knowledge management system and policies and procedures

21. Assume you have identified your stakeholders and are preparing your stakeholder engagement plan. You are fortunate that your team is a collocated team as you are working on an internal project to reorganize your IT Department, so it is focused more on its customers. The project sponsor is the Chief Operating Officer, and the IT Department Director was surprised as she thought all was well. However, you notice when planning meetings are held, the Chief Financial Officer never attends. You feel since IT affects the entire company, all the senior leaders need some type of involvement. You therefore feel the Chief Financial Officer may be—

    a. Resistant
    b. Unaware
    c. Uninterested
    d. Satisfied

22. Assume your stakeholder engagement plan has been approved. You now are working with your team to promote stakeholder engagement on your project. You explain in a team meeting its benefit is to—

    a. Clarify and resolve identified issues
    b. Meet stakeholder needs and expectations
    c. Obtain their continued commitment to the project
    d. Increase support and minimize resistance

23. The stakeholder register should not be prepared only one time, but it should be updated regularly especially if—

    a. The stakeholder is not an active participant
    b. The stakeholder has new information
    c. The stakeholder does not read status updates
    d. The stakeholder leads a corporate reorganization

24. Working to foster stakeholder engagement, as the project manager, you know interpersonal and team skills are needed. An example of a key interpersonal and team skill in Manage Stakeholder Engagement is—

    a. Facilitating consensus
    b. Influencing people
    c. Resolving conflicts
    d. Resolving issues

25. Stakeholder engagement must be monitored on a continuous basis for it to be effective. You realize a number of project documents can be useful for you as a project manager. An example is—

    a. Technical performance measures
    b. Change log
    c. Risk register
    d. Start and finish dates of schedule activities

26. Expert judgment is a tool and technique in Identify Stakeholders. To use it effectively on your project, you should—

    a. Use a focus group
    b. Review documentation
    c. Hold brainstorming sessions with your team
    d. Use knowledge from team member contributions

27. Enterprise environmental factors are useful in many projects. In Plan Stakeholder Engagement an example of one of particular importance is—

a. Lessons learned
b. Templates from previous stakeholder engagement plans from successful projects
c. Benchmarking
d. Organizational culture

28. Stakeholder engagement involves more than improving communications and requires more than just managing a team. This is because it—

a. Contains detailed plans to ensure all stakeholder requirements are satisfied
b. Describes how to best meet human resource requirements
c. Identifies the strategies and actions to promote stakeholder involvement
d. Focuses on active participation of all stakeholders even the resistors to the project

29. Assume you are beginning your project to develop a series of residential condominiums in your city and are identifying possible stakeholders. A key organizational process asset you can review is—

a. Organizational culture
b. Organizational standards
c. Stakeholder registers from previous projects
d. Local trends

30. One way to develop an understanding of major project stakeholders to exchange and analyze project information about roles and interests is to—

a. Conduct interviews
b. Hold profile analysis meetings
c. Use questionnaires and surveys
d. Conduct a stakeholder analysis

31. Assume you are managing a project to implement a Software as a Service medical record system in your ophthalmologist's office. You have been working to identify your stakeholders to then make sure everyone is committed to it as some people have been working in this office for more than 20 years and are comfortable with the current approach. At this point, you used document analysis, which includes—

a. Lessons learned from previous projects
b. Whether the stakeholder is a supporter, is neutral, or is resistant
c. Potential influence in the project
d. Organization position

32. Having prepared stakeholder engagement plans on previous projects, you know it is positive to review the project management plan because it—

    a. Provides information as to how to plan appropriate ways to engage stakeholders
    b. Contains information useful to ensure the stakeholder engagement plan is aligned with the organization's culture
    c. Helps to determine the best options to support an adaptive process for stakeholder engagement
    d. Contains a resource management plan

33. Assume you have performed your stakeholder analysis and now are working to enhance it with a stakeholder engagement assessment matrix. Such a matrix shows the stakeholder's current engagement level. These data help—

    a. The project manager to prepare the stakeholder engagement plan
    b. The project manager to prepare the stakeholder engagement strategy
    c. The project manager to prepare the stakeholder inventory
    d. The project team to expand the stakeholder risk register

34. A trend and emerging practice in Project Stakeholder Engagement is to—

    a. Consider the negative value of stakeholder engagement
    b. Recognize stakeholders have different levels of interest in the project at different times
    c. Interview the stakeholders with high power and high influence
    d. Involve as many stakeholders as possible in planning stakeholder management

35. Working on your project to design and construct five new bridges for your City, you are striving to actively engage the stakeholders on your project, especially those who will be inconvenienced by the project and have indicated they do not support it. You decided to—

    a. Establish issue management procedures
    b. Work with the positive stakeholders to determine processes to use in each life cycle
    c. Use co-creation
    d. Provide guidance as to how to best involve stakeholders in the project

36. An input to consider for the Monitor Stakeholder Engagement process is—

    a. Budget
    b. Ethics corporate policies
    c. Historical information
    d. Number of defects

37. As a result of the Monitor Stakeholder Engagement process, you realize even though this process is under way from the beginning until the closing phase that you have identified the root cause of some issues you have faced in monitoring stakeholders' expectations. You should therefore—

    a. Review them with your Governance Board
    b. Revise and reissue your stakeholder management plan
    c. Prepare a change request
    d. Update the lessons learned documentation

38. Identifying strategies to promote productive involvement of stakeholders is useful to the project manager as he or she works with stakeholders. It should be documented as part of the—

    a. Stakeholder register
    b. Stakeholder engagement strategy
    c. Stakeholder engagement plan
    d. Stakeholder engagement assessment matrix

39. A number of organizational process assets are useful as inputs to the Manage Stakeholder Engagement process. An example is—

    a. Project reports
    b. Established communications channels
    c. Project records
    d. Organizational communications requirements

40. Multi-criteria decision analysis and voting are examples of—

    a. Project reports as an input to Monitor Stakeholder Engagement
    b. Work performance information as an output of Monitor Stakeholder Engagement
    c. Tools and techniques used in Monitor Stakeholder Engagement
    d. Updates from the Plan Stakeholder Engagement process

Answer Sheet

1.	a	b	c	d		21.	a	b	c	d
2.	a	b	c	d		22.	a	b	c	d
3.	a	b	c	d		23.	a	b	c	d
4.	a	b	c	d		24.	a	b	c	d
5.	a	b	c	d		25.	a	b	c	d
6.	a	b	c	d		26.	a	b	c	d
7.	a	b	c	d		27.	a	b	c	d
8.	a	b	c	d		28.	a	b	c	d
9.	a	b	c	d		29.	a	b	c	d
10.	a	b	c	d		30.	a	b	c	d
11.	a	b	c	d		31.	a	b	c	d
12.	a	b	c	d		32.	a	b	c	d
13.	a	b	c	d		33.	a	b	c	d
14.	a	b	c	d		34.	a	b	c	d
15.	a	b	c	d		35.	a	b	c	d
16.	a	b	c	d		36.	a	b	c	d
17.	a	b	c	d		37.	a	b	c	d
18.	a	b	c	d		38.	a	b	c	d
19.	a	b	c	d		39.	a	b	c	d
20.	a	b	c	d		40.	a	b	c	d

Answer Key

1. d. Recognize that roles and responsibilities may overlap but practice a policy of 'no surprises' and inform your stakeholders about any upcoming meetings

    Meetings are a tool and technique in Monitor Stakeholder Engagement. They include status meetings, standup meetings, retrospectives, or others as stated in the stakeholder engagement plan. They help to monitor stakeholder engagement levels. The intent of inviting stakeholders to meetings or reviews or posting project artifacts in public space is to surface quickly any misalignments, dependencies, or other issues in the project. It is a concept used in agile methods to promote aggressive transparency and to accelerate the flow of information sharing. [Monitoring and Controlling]

    PMI®, *PMBOK® Guide*, 2017, 506, 535
    PMI® *PMP Examination Content Outline*, 2015, Monitoring and Controlling, 9, Task 1

2. b. Consult the enterprise environmental factors

    The key words in this question are the team members are located in three different continents. The geographic distribution of facilities and resources are enterprise environmental factors to consider in the Monitor Stakeholder Engagement process. [Monitoring and Controlling]

    PMI®, *PMBOK® Guide*, 2017, 533
    PMI® *PMP Examination Content Outline*, 2015, Monitoring and Controlling, 9, Task 1

3. c. Comparing upward, downward, outward, and sideward approaches

    Identifying and analyzing the stakeholders helps to classify them to develop a strategy to meet their expectations throughout the project. In this approach, you are analyzing them in terms of upward approaches to senior managers, customer, sponsors, and the steering committee; downward to the team or SMEs; outward to stakeholders and groups outside of the team such as suppliers, the government, public, end users, or regulators; and sideward to your peers or middle managers who also want the same scarce resources you need on your project. [Initiating]

    PMI®, *PMBOK® Guide*, 2017, 513
    PMI® *PMP Examination Content Outline*, 2015, Initiating, 5, Task 3

4. a. Conduct a stakeholder analysis to collect information

   Stakeholder analysis is used to result in a stakeholder list and information on the stakeholders such as their positions in the organization, roles on the project, stakes or expectations, attitudes or levels of support, and their interest in information about the project. It is a data analysis tool and technique used in Identify Stakeholders. [Initiating]

   PMI®, *PMBOK® Guide*, 2017, 512
   PMI® *PMP Examination Content Outline,* 2015, Initiating, 5, Task 3

5. b. Provide a clear plan that is actionable to interact effectively with stakeholders

   The Plan Stakeholder Engagement process develops appropriate management strategies to involve stakeholders based on their interests, needs, expectations and potential impact on the project. The key benefit of this process is to have a plan that is clear and actionable to interact with them. [Planning]

   PMI®, *PMBOK® Guide*, 2017, 516
   PMI® *PMP Examination Content Outline,* 2015, Planning, 8, Task 13

6. b. The stakeholder register

   The stakeholder register, along with the issue log, change log, and lessons learned register, are project documents that may need updates in the Manage Stakeholder Engagement process. [Executing]

   PMI®, *PMBOK® Guide*, 2017, 529
   PMI® *PMP Examination Content Outline,* 2015, Executing, 8, Task 7

7. c. An output from the Manage Stakeholder Engagement process

   Issues logs are an output of this process, as issues are expected in this process. The log is updated to reflect an update to or the development of an issue log entry. [Executing]

   PMI®, *PMBOK® Guide*, 2017, 629
   PMI® *PMP Examination Content Outline,* 2015, Executing, 8, Task 7

8. d. Determine how changes will be monitored and controlled

The key words in this question are monitoring overall project stakeholder relations, which indicates you are in the Monitor Stakeholder Engagement process. Its purpose is to monitor these relationships and tailor strategies to engage stakeholders as you modify engagement strategies and plans. As you work in managing stakeholder engagement, you should review corporate policies and procedures for issue, risk, change, and data management, one of the organizational process assets that are inputs to this process. You also may need a change request with this change even though it is mandated by the CEO, an output of this process, as it describes the preventive and corrective actions [as in this question], to improve the current level of stakeholder engagement. [Monitoring and Controlling]

PMI®, *PMBOK® Guide*, 2017, 530, 533, 535
PMI® *PMP Examination Content Outline*, 2015, Monitoring and Controlling, 9, Task 1

9. b. Show gaps between current and desired levels of engagement

The stakeholder engagement assessment matrix is a data representation tool used as a tool and technique in Plan Stakeholder Management. The purpose of the matrix is to show gaps between current and desired engagement levels required for successful project delivery. This engagement level classifies stakeholders to show if they are unaware, resistant, neutral, supportive, or leading. It also is useful to determine the level of communications needed to engage stakeholders. [Planning]

PMI®, *PMBOK® Guide*, 2017, 522–523
PMI® *PMP Examination Content Outline*, 2015, Planning, 7, Task 13

10. c. When the organization structure changes

Other trigger conditions are when it is time to start a new project phase; industry changes; when new people or groups become stakeholders, current stakeholders are no longer involved, or their importance to project success changes; and when outputs of other process areas such as change management, risk management, or issue management require a review of stakeholder engagement strategies. [Planning]

PMI®, *PMBOK® Guide*, 2017, 518
PMI® *PMP Examination Content Outline*, 2015, Planning, 8, Task 13

11. c. Identifies the appropriate focus for each stakeholder or a group of stakeholders

   The Identify Stakeholder process has a number of purposes. It identifies stakeholders regularly. It analyzes and documents relevant information concerning their interests, involvement, interdependencies, influence, and potential impact on project success. Its key benefit is to allow the project manager and the team to identify the appropriate focus for each stakeholder or stakeholder groups. [Initiating]

   PMI®, *PMBOK® Guide*, 2017, 507
   PMI® *PMP Examination Content Outline*, 2015, Initiating, 5, Task 3

12. c. Brain writing

   In data gathering, questionnaires and surveys, brainstorming, and brain writing are techniques to consider. Brain writing is a form of brainstorming, as is the Nominal Group Technique, another possible answer, but it focuses on giving the participants time to consider the questions in advance of the brainstorming meeting or group creativity meeting. Then, people can speak up during the session, whether it is virtual or co-located, as they have had time to think about the answers. it helps avoid one or two people from dominating the session and enables more equal participation. [Initiating]

   PMI®, *PMBOK® Guide*, 2017, 511
   PMI® *PMP Examination Content Outline*, 2015, Initiating, 5, Task 3

13. b. Use root-cause analysis

   Root-cause analysis is a data analysis approach in Plan Stakeholder Engagement. It purpose is to identify the reasons for the level of support of the stakeholders in the project and then use it to determine the most appropriate strategy to improve their level of engagement. Other data analysis techniques are assumption and constraint analysis. [Planning]

   PMI®, *PMBOK® Guide*, 2017, 521
   PMI® *PMP Examination Content Outline*, 2015, Planning, 8, Task 13

14. a. Managing stakeholder expectations by using negotiation and communications

A key activity in Manage Stakeholder Engagement is to manage stakeholder expectations through negotiation and communications, which can help ensure project goals are achieved. This process focuses on communicating with stakeholders and working with them to meet their needs or expectations, addressing issues as they occur, and fostering appropriate involvement. [Executing]

PMI®, *PMBOK® Guide*, 2017, 523–524
PMI® *PMP Examination Content Outline*, 2015, Executing, 8, Task 7

15. d. Status of stakeholder engagement

Work performance information includes information about the status of stakeholder engagement. It includes the level of current project support compared to the desired level of engagement, which is defined in tools such as the stakeholder engagement assessment matrix and the stakeholder cube. [Monitoring and Controlling]

PMI®, *PMBOK® Guide*, 2017, 535
PMI® *PMP Examination Content Outline*, 2015, Monitoring and Controlling, 9, Task 1

16. c. Work actively with members of your Governance Board

New product development organizations are noted for setting up Governance Boards to oversee projects. Additionally, in this situation, it is a long project that is important to the company. Project governance refers to the framework, functions, and processes to guide project activities, regardless if the project is for a product, service, or result. Its objective is to meet organizational, strategic, and operational goals. Recognize there is no one governance framework that supports all projects. [Executing]

PMI®, *PMBOK® Guide*, 2017, 44
PMI® *PMP Examination Content Outline*, 2015, Executing, 8, Task 7

17. d. You need to continually identify project stakeholders

Stakeholder identification is a continual process throughout the project life cycle. The legal department often is overlooked, but it is a significant stakeholder, and in this situation, delays resulted. Significant expenses often are due to legal requirements that must be met before the project can be completed, or the project scope is delivered. As an output of Identify Stakeholders, the stakeholder register should be consulted and updated; stakeholders may change, or new stakeholders will be identified. Stakeholder identification is performed throughout the project. [Initiating]

PMI®, *PMBOK® Guide*, 2017, 31, 507
PMI® *PMP Examination Content Outline*, 2015, Initiating, 5, Task 3

18. d. Balance the interests of these negative stakeholders and meet with them before Tuesday's meeting

Overlooking negative stakeholders' interests can result in an increased likelihood of failures, delays, or other negative consequences to projects. The project manager must control stakeholder engagement, which can be difficult since they often have different or competing objectives. The key benefit of the Monitor Stakeholder Engagement process is to maintain or increase the efficiencies of stakeholder engagement activities as the project evolves and the project environment changes. Therefore, it is necessary to monitor stakeholder relationships and adjust strategies and plans to engage stakeholders. [Monitoring and Controlling]

PMI®, *PMBOK® Guide*, 2017, 530
PMI® *PMP Examination Content Outline*, 2015, Monitoring and Controlling, 9, Task 1

19. a. Stakeholders' power, urgency and legitimacy are used

In the salience model, stakeholders are described in classes based on their power or level of authority or ability to influence the project, urgency or need for immediate action, and legitimacy or their involvement. It is an approach used for data representation along with power/interest, power/influence, and influence/impact grids, the stakeholder cube, prioritization, and meetings. Data representation is a tool and technique in Identify Stakeholders to categorize stakeholders using various methods. [Initiating]

PMI®, *PMBOK® Guide*, 2017, 512–513
PMI® *PMP Examination Content Outline*, 2015, Initiating, 5, Task 3

20. c. Lessons learned repository

Of the answers provided, the lessons learned repository is the only organizational process asset. It contains information about stakeholder preferences, actions, and involvement; therefore, it is useful to consult as the stakeholder engagement plan is prepared. [Planning]

PMI®, *PMBOK® Guide*, 2017, 520
PMI® *PMP Examination Content Outline*, 2015, Planning, 7, Task 13

21. a. Resistant

Since the Chief Financial Officer has financial responsibility for all of the company's work, in preparing a stakeholder engagement plan, he or she probably is aware of this project and probably is resistant to change. This resistance may be because of changes that will occur based on the project's work and its outcomes and will not be supportive. It is a data representation tool and technique in Plan Stakeholder Engagement. It is a stakeholder engagement assessment matrix used to compare stakeholders' current engagement levels and desired levels for success. Other ways to classify stakeholder engagement levels are unaware, neutral, supportive, or leading. [Planning]

PMI®, *PMBOK® Guide*, 2017, 521
PMI® *PMP Examination Content Outline*, 2015, Planning, 7, Task 13

22. d. Increase support and minimize resistance

The benefit of the Manage Stakeholder Engagement process is to allow the project manager to increase support and minimize resistance from stakeholders. In this process, the purpose is to communicate with stakeholders to meet needs and expectations, address issues, and foster active involvement. [Executing]

PMI®, *PMBOK® Guide*, 2017, 523
PMI® *PMP Examination Content Outline*, 2015, Executing, 8, Task 7

23. b. The stakeholder has new information

Project document updates are an output to the Manage Stakeholder Engagement process. These updates involve the stakeholder register. It should be updated because of new information provided to stakeholders on resolved issues, approved changes, and overall project status. [Executing]

PMI®, *PMBOK® Guide*, 2017, 529
PMI® *PMP Examination Content Outline*, 2015, Executing, 8, Task 7

24. c. Resolving conflicts

Conflicts are common on projects and between stakeholders and the project manager must resolve them in a timely way. Other interpersonal and team skills useful in Manage Stakeholder Engagement are cultural awareness, negotiation, observation/conversation, and political awareness. [Executing]

PMI®, *PMBOK® Guide*, 2017, 527
PMI® *PMP Examination Content Outline*, 2015, Executing, 8, Task 7

25. c. Risk register

A project document to review as an input to the Monitor Stakeholder Engagement process is the risk register. It is useful since it contains the project's identified risks, which can include stakeholder engagement and interaction, categorization, and potential responses. Other project documents to consider are the issue log, project communications, lessons learned register, and the stakeholder register. [Monitoring and Controlling]

PMI®, *PMBOK® Guide*, 2017, 532
PMI® *PMP Examination Content Outline*, 2015, Monitoring and Controlling, 9, Task 1

26. d. Use knowledge from team member contributions

Expert judgment in the Identify Stakeholder process is considered from people or groups with specialized knowledge on a variety of topics. Team members bring technical expertise to the project, and their contributions should not be overlooked. The project manager then should be familiar with the team members' knowledge, skills, and competencies. Other topics of interest in expert judgment are understanding the organization's politics and power structures, knowledge of the organization's environment and culture, and knowledge of the project's industry. [Initiating]

PMI®, *PMBOK® Guide*, 2017, 511
PMI® *PMP Examination Content Outline*, 2015, Initiating, 5, Task 3

27. d. Organizational culture

    Enterprise environmental factors are inputs to Plan Stakeholder Management because the engagement of stakeholders should be adapted to the project environment. This means organizational culture is of particular importance, along with the political, climate, and governance framework. These three help determine the best option to support a better and more adaptive approach for planning stakeholder engagement. [Planning]

    PMI®, *PMBOK® Guide*, 2017, 519
    PMI® *PMP Examination Content Outline*, 2015, Planning, 7, Task 13

28. c. Identifies the strategies and actions to promote stakeholder involvement

    Plan Stakeholder Engagement has as its output the stakeholder engagement plan. It is necessary to identify these strategies and actions such that there is productive involvement from stakeholders in decision making and execution. It may include specific strategies or approaches to use based on the project's needs and stakeholder expectations. [Planning]

    PMI®, *PMBOK® Guide*, 2017, 522
    PMI® *PMP Examination Content Outline*, 2015, Planning, 7, Task 13

29. c. Stakeholder registers from previous projects

    Stakeholder registers from previous projects is the only answer that is an organizational process asset. Other organizational process assets are a lesson learned repository that has information about preferences, actions, and involvement of stakeholders and a stakeholder register with instructions. [Initiating]

    PMI®, *PMBOK® Guide*, 2017, 510
    PMI® *PMP Examination Content Outline*, 2015, Initiating, 5, Task 3

30. d. Conduct a stakeholder analysis

    A stakeholder analysis is a data analysis tool and technique the Identify Stakeholders process. In addition to roles and interests, it results in information on positions in the organization, "stakes", and attitudes. [Initiating]

    PMI®, *PMBOK® Guide*, 2017, 512
    PMI® *PMP Examination Content Outline*, 2015, Initiating, 5, Task 3

31. a. Lessons learned from previous projects

    Document analysis is the other data analysis technique in Identify Stakeholders. The other is to assess available project documentation. [Initiating]

    PMI®, *PMBOK® Guide*, 2017, 512
    PMI® *PMP Examination Content Outline*, 2015, Initiating, 5, Task 3

32. d. Contains a resource management plan

    In Plan Stakeholder Engagement, project management plan components to consider are the resource management, communications management, and risk management plan. The resource management plan is helpful since it has information on the team and other stakeholders' roles and responsibilities based on the stakeholders listed in the stakeholder register. [Planning]

    PMI®, *PMBOK® Guide*, 2017, 518
    PMI® *PMP Examination Content Outline*, 2015, Planning, 7, Task 13

33. a. The project manager to prepare the stakeholder engagement plan

    The stakeholder engagement assessment matrix shows the stakeholders current engagement in the project and the desired engagement level needed for project success. Stakeholders in this matrix can be classified as unaware, resistant, neutral, supportive, or leading. [Planning]

    PMI®, *PMBOK® Guide*, 2017, 521
    PMI® *PMP Examination Content Outline*, 2015, Planning, 7, Task 13

34. a. Consider the negative value of stakeholder engagement

    There are several trends and emerging practices in project stakeholder engagement. One is to capture both the positive and the negative value of stakeholder engagement. Positive value considers benefits from higher levels of stakeholder support. Negative value is derived by measuring the true costs of not engaging stakeholders effectively as by not doing so, it could lead to product recalls or loss of the organization's and/or the project's reputation. [Executing]

    PMI®, *PMBOK® Guide*, 2017, 505
    PMI® *PMP Examination Content Outline*, 2015, Executing, 8, Task 7

35. c. Use co-creation

   Another emerging trend and practice in Project Stakeholder Engagement is co-creation. It is an approach to place greater emphasis on including the affected stakeholders in the team as partners, which fits the situation in this question. It is used to consult with the stakeholders who are the most affected by the work or the project's outcomes. [Executing]

   PMI®, *PMBOK® Guide*, 2017, 505
   PMI® *PMP Examination Content Outline*, 2015, Executing, 8, Task 7

36. b. Ethics corporate policies

   A number of organizational process assets are useful to consider in Monitor Stakeholder Engagement. One of importance to any project is corporate policies and procedures for ethics. The organization's code of ethics or code of conduct should be considered. Other corporate policies and procedures involve use of social media; security; and issue, risk, change and data management. [Monitoring and Controlling]

   PMI®, *PMBOK® Guide*, 2017, 533
   PMI® *PMP Examination Content Outline*, 2015, Monitoring and Controlling, 9, Task 1

37. c. Prepare a change request

   A change request is an output of the Monitor Stakeholder Engagement process. This change request may involve corrective and preventive actions taken as you analyzed the root cause in this situation in the question. Other outputs are work performance information and updates to the resource management, communications management, and stakeholder engagement plans. [Monitoring and Controlling]

   PMI®, *PMBOK® Guide*, 2017, 535
   PMI® *PMP Examination Content Outline*, 2015, Monitoring and Controlling, 9, Task 1

38. c. Stakeholder engagement plan

   The stakeholder engagement plan is the output of Plan Stakeholder Management. It is a component of the project management plan. Its purpose is to identify the strategies or actions to promote productive involvement of stakeholders in decision making and project execution. It may be limited to specific stakeholders or stakeholder groups. [Planning]

   PMI®, *PMBOK® Guide*, 2017, 522
   PMI® *PMP Examination Content Outline*, 2015, Planning, 7, Task 13

39. d. Organizational communications requirements

Many organizations have established communications requirements such as who must approve information that is communicated to certain stakeholders whether they are internal or external to the project. If your organization has such policies, you should review them as you Manage Stakeholder Engagement. [Executing]

PMI®, *PMBOK® Guide*, 2017, 526
PMI® *PMP Examination Content Outline*, 2015, Executing, 8, Task 7

40. c. Tools and techniques used in Monitor Stakeholder Engagement

In Monitor Stakeholder Engagement, they are examples of decision-making techniques used in Monitor Stakeholder Engagement. Multi-criteria decision analysis can be used to prioritize and weight the criteria to be used for successful stakeholder engagement and to identify the best choice. Voting also can be used to select the best response if there is a variance in stakeholder engagement. [Monitoring and Controlling]

PMI®, *PMBOK® Guide*, 2017, 534
PMI® *PMP Examination Content Outline*, 2015, Monitoring and Controlling, 9, Task 1

Practice Test

This practice test is designed to simulate PMI®'s 200-question PMP® certification exam.

INSTRUCTIONS: Note the most suitable answer for each multiple-choice question in the appropriate space on the answer sheet.

1. You recently took over a relatively new project expected to last another seven years. The previous project manager completed most of the WBS. When you begin to define the project activities, you realize that the WBS work packages expected to occur in the next year are planned in detail, but the work packages for later in the future (three years or more) are not planned with much detail, if any detail at all. You determine—

   a. It is a major problem. The WBS is incomplete and you need to redefine the project scope to complete the project schedule.
   b. It is a problem that must be resolved quickly. The previous project manager was not done with the WBS, and you must stop the project to complete the WBS in sufficient detail.
   c. It is not a problem at this time. The previous project manager was using the rolling wave planning technique, so you are able to continue defining the activities.
   d. It is not a problem at this time. You can only plan what you know. You plan to communicate to the project sponsor that the WBS is not sufficient to plan the whole project and that the sponsor can worry about the details.

2. You are managing a project to transform your organization into one that is more customer centric as your executives realize they no longer can assume existing customers will stay with your firm given offerings by the competition. You and your team identified over 300 stakeholders and classified them. You realize 45 are influential but negative to the project. You then prepared your stakeholder engagement plan. Now, you find you are addressing concerns from these stakeholders plus others, and it consumes most of your time. However, it is important because—

   a. You want to increase their support and minimize resistance
   b. The stakeholders only wish to work with you
   c. You need committed stakeholders to ensure your project continues
   d. You realize your project has the greatest chance for success if it operates in a projectized environment, and stakeholder support is paramount

3. On your project to construct a new runway for your City's airport, you are in the process of selecting sellers for various parts of this project. You have conducted your make-or-buy analysis and have issued Requests for Proposals. Before accepting a proposal, you decide to—

   a. Review invoicing and payment information
   b. Set up a multi-disciplinary review team
   c. Use analytical techniques
   d. Establish a system to control claims

4. Requirements always are a concern to ensure stakeholder needs are met. Your organizational leaders even have noted the importance of capturing the right requirements to help ensure project success. They have set up a business analysis group, and everyone in it is certificated in business analysis. Since your project is one that has captured your senior manager's attention, the importance of collecting the right requirements and avoiding scope creep is emphasized by your sponsor and other key stakeholders. A business analyst is working with you and your team in collaborative way. As you wok to collect requirements you first focus on—

   a. Involving as many key stakeholders as possible through interviews and questionnaires
   b. Recognizing the appropriate level of service requirements
   c. Determining problems and identifying business needs
   d. Performing the requirements activities quickly

5. The project life cycle, management reviews, and the development approach in the Develop Project Management Plan process are—

   a. Enterprise environmental factors
   b. Organizational process assets
   c. Part of the project management plan
   d. Part of the organization's management practices

6. You are managing a project that has five subcontractors. You must monitor contract performance, make payments, and manage provider interfaces. One subcontractor submitted a change request to expand the scope of its work. You decided to award a contract modification based on a review of this request. All these activities are part of—

   a. Control Procurements
   b. Conduct Procurements
   c. Form Contract
   d. Configuration Management

7. Your project management plan has been approved, and since your company follows a stage-gate approach, you are now in the executing phase. You have collected a lot of data, and these data are viewed as—

   a. Ones set forth in the PMIS
   b. The lowest level of detail to derive information
   c. Ones recommended through a survey of your stakeholders to assess their key communications requirements
   d. A major part of the project management plan

8. Assume you are managing a small project with a budget of $50,000 to redesign how training is conducted in your organization. Your team has three people who work on the project part time and are subject matter experts. Your project charter has been approved, and now you are working on your project management plan. You know a best practice is to have a kick-off meeting so you—

   a. Use it to involve your team in the planning process
   b. Use it when the plan is approved, and you move into the executing phase
   c. Use it to discuss team skills and roles and responsibilities
   d. Use it at the beginning of each phase in the project for team building

9. While working as the project manager on a new project to improve overall ease of use in the development of a railroad switching station, you have decided to add a subject matter expert who specializes in ergonomics to your team. She has decided to observe the existing approach as you and your team work to define requirements for the new system. This method is also called—

   a. Mentoring
   b. Coaching
   c. Job shadowing
   d. User experimentation

10. In addition to providing support to the project, Manage Quality also provides an umbrella for—

   a. Plan-do-check-act
   b. Improvement of quality planning
   c. Project management maturity
   d. Work performance information

11. As you manage the railroad switching station project, you are concerned that the business analyst who was responsible for preparing the WBS may have overlooked some parts of the project. In order to see if the WBS requires enhancements you decide to—

   a. Perform a cause-and-effect diagram
   b. Meet with your sponsor
   c. Use an affinity diagram
   d. Review the accompanying WBS Dictionary with a member of the PMO

12. Assume that you are managing a project in which 80% has been outsourced to various sellers. In terms of contract administration, a key aspect is—

   a. Having a contract administrator assigned to your core team and reports to you
   b. Ensuring there are few, if any, claims
   c. Considering the sellers as members of the project team
   d. Managing the interfaces among the sellers

13. Manual and automated tools can support configuration management and change management. The use of the specific tools supports project stakeholder needs, which include organizational and environmental consideration and/or constraints. Since your project involves the development of flying cars in congested urban areas, you believe you should use both configuration and change control. In configuration control, your last step is to—

    a. Determine specific responsibilities
    b. List the approved configuration identification
    c. Perform verification and audit on each configuration item
    d. Ensure the composition of a project's configuration items is correct

14. You need to outsource the testing function of your project. Your subcontracts department informed you that the following document must be prepared before conducting the procurement—

    a. Make-or-buy analysis
    b. Procurement management plan
    c. Evaluation methodology
    d. Contract terms and conditions

15. Assume you are working to Monitor and Control Project Work. A useful tool and technique is—

    a. Monitoring and reporting methods
    b. Requirements traceability matrix
    c. Variance analysis
    d. Influence diagrams

16. You are pleased to be managing a project to enhance and upgrade the landfill system in your city. Nothing has changed since it was first created more than 20 years ago. You prepared your business case along with your sponsor, and the project is authorized because of a—

    a. Sustainability
    b. Customer request
    c. Social need
    d. Ecological impact

17. To identify inefficient and ineffective policies, processes, and procedures in use on a project, you should conduct—

    a. An inspection
    b. A process analysis
    c. Benchmarking
    d. A quality audit

18. Your project management office implemented a project management methodology that emphasizes the importance of integrated change control. It states that change requests may include—

   a. Indirect changes
   b. Defect repairs
   c. Informal from executives
   d. Ones that do not impact baselines

19. Since projects involve change, change requests then will be needed. Every documented change request must be approved by a responsible person, who is—

   a. The project manager or the sponsor
   b. A member of the CCB
   c. Identified in the change management plan
   d. Identified in the project management plan

20. A number of tools and techniques are helpful in the Perform Integrated Change Control process. You have a CCB on your project, and with it, you plan to use—

   a. Configuration management software
   b. A project management information system
   c. Project status review meetings
   d. Change control meetings

21. Having worked previously as a software project manager, you were pleased to be appointed as the project manager for a new systems integration project designed to replace the existing air traffic control system in your country. You found a requirements traceability matrix to be helpful on software projects, so you decided to use it on this systems integration project. Using such a matrix helps to ensure that each requirement—

   a. Adds quality and supports the organization's quality policy
   b. Adds business value as it links to business and project objectives
   c. Sets forth the level of service, performance, safety, security, and compliance
   d. Shows the impact to other organizational areas and to entities outside of the performing organization

22. You are working on a complex project in the medical device field. You have about 55 internal stakeholders and a very large number of external stakeholders with the regulatory and consumer interest groups along with the medical profession. You need to be aware of their risk tolerances as they are examples in the Direct and Manage Project Work process of—

    a. The need to review the stakeholder management plan early in the process
    b. Enterprise environmental factors
    c. Organizational process assets
    d. The importance of classifying stakeholders as to interest and influence

23. Working in the Perform Integrated Change Control process, you want to study the impact of the change on the project's scope. Therefore, you need to consider the—

    a. Scope change control system
    b. Basis of estimates
    c. Requirements traceability matrix
    d. Performance management baseline

24. Your company has over the past two years changed its methodologies and now is using agile. You are working to Manage Quality on your project. Therefore, your focus is on—

    a. Using Pareto analysis
    b. Concentrating on small batch systems
    c. Analyzing cost-benefit periodically
    d. Determining quality metrics

25. You are working to validate the scope of your project to ensure it meets acceptance criteria. You also are working on Control Quality. Therefore, you want to—

    a. Ensure processes are performed completely in parallel
    b. Both processes only use inspection as a tool and technique
    c. Ensure Control Quality is concerned with meeting quality requirements for the deliverables
    d. Ensure Validate Scope typically precedes Control Quality

26. There are a number of activities for administrative closure of the project or a phase. The first one is—

    a. Transitioning the product or service to end users, the customer, or an operations group
    b. Measuring the completeness of the project's deliverables against the requirements
    c. Making sure all the activities necessary to satisfy exit criteria for a phase or the entire project are followed
    d. Holding a lessons-learned session with interested stakeholders

27. You are in the process of performing managing quality on your project's product and find that some requirements are not as complete as they should be, which causes rework and adds costs to your overall project. One approach to decide your next steps is—

    a. Determining the cost of quality
    b. Using a checklist
    c. Finding solutions for these issues and other challenges
    d. Comparing the cost of rework to the life-cycle costs of the project

28. Assume you are working on a complex project, with team members located in four countries. Your company is striving to be the first to market with a new medical device that if a patient requires a stent, he or she will never require another one. Your project is in the top three in the company's portfolio, and executives are reviewing it closely. You have a number of unknown unknowns on this project, which means—

    a. You are conducting assumptions analysis using a subject matter expert.
    b. You are holding bi-weekly reviews to identify emergent risks.
    c. You are analyzing variability risks.
    d. You have added an expert in risk management to your team.

29. Assume your organization has changed its project management methodology so it is focused on an agile/adaptive environment. This change took time to implement since people tend to resist change, but now some are recognizing its value. You decided that since you had your PMP to become certified as well in agile. Now you are using an agile approach on your project. In this approach, you—

    a. Need to spend less time defining requirements early on your project
    b. Use affinity diagrams to capture a variety of requirements and then categorize them in the later stages
    c. Use business analyst in which this person has complete requirements responsibility
    d. Apply approaches such as ensuring requirements are stable from the start

30. You are executing your organizational transformation project to move toward a managing-by-project culture. To do so effectively, you want to connect people to people, so they can work together and create new knowledge as this project continues. Your goal is to share tacit knowledge and also integrate the knowledge of your diverse team. It is an innovative and complex project. To help you in this area, you decide to use—

    a. Lessons learned register
    b. Work shadowing
    c. Employee development and training records
    d. Organizational learning

31. An approach to provide insight into the health of the project and to identify any areas that require special attention is to—

    a. Conduct periodic status reviews
    b. Prepare regular status and progress reports
    c. Prepare forecasts of the project's future
    d. Continuously monitor the project

32. Although your company's project life cycle does not mandate when a project review should be conducted, you believe it is important to review performance at the conclusion of each phase. The objective of such a review is to—

    a. Determine how many resources are required to complete the project according to the project baseline
    b. Adjust the schedule and cost baselines based on past performance
    c. Obtain customer acceptance of project deliverables
    d. Determine whether the project should continue to the next phase

33. In the performing project stage of team development, it now is time for the project manager to—

    a. Evaluate performance
    b. Empower the team
    c. Celebrate success
    d. Improve trust

34. Assume that your actual costs are $1,000; your planned value is $1,200; and your earned value is $1,500. Based on these data, what can be determined regarding your schedule variance?

    a. At −$300, the physical progress is being accomplished at a slower rate than is planned, indicating an unfavorable situation.
    b. At +$300, the situation is favorable, as physical progress is being accomplished ahead of your plan.
    c. At +$500, the situation is favorable, as physical progress is being accomplished at a lower cost than was forecasted.
    d. At −$300, you have a behind-schedule condition, and your critical path has slipped.

35. You are concerned because when you did your stakeholder engagement assessment matrix, one of the stakeholders you classified as leading seems to now lack interest in your project. This stakeholder, your Chief Financial Officer, is someone you consider critical to success. You have been monitoring stakeholder engagement throughout the project. You realize you need to meet with this stakeholder personally and have a meeting scheduled. You want to see if the Chief Financial Officer feels you need to modify your stakeholder engagement strategies and plan. During this meeting you plan to—

    a. Focus on whether he feels too much or not enough information is provided
    b. Ask open-ended questions and use active listening
    c. Inform him of the project's cost status to ensure he still will provide the funds needed in your cost management plan
    d. Make a presentation about the entire project

36. The CPI on your project is 0.44, which means that you should—

    a. Place emphasis on improving the timeliness of the physical progress
    b. Reassess the life-cycle costs of your product, including the length of the life–cycle phase
    c. Place emphasis on improving the productivity by which work was being performed
    d. Recognize that your original estimates were fundamentally flawed, and your project is in an atypical situation

37. Assume your organization is adopting the use of agile, but before it is used on all projects, you are testing its use on your project. Even though agile has different methods, you must prepare a cost estimate. The best approach to use is—

   a. Agile release planning
   b. An analogous estimate
   c. A high-level cost forecast
   d. Guidelines from the project's governance committee

38. Which of the following tools and techniques is used in the Close Project or Phase process?

   a. Project management methodology
   b. Work performance information
   c. Expert judgment
   d. Project management information system

39. Assume you have prepared your cost estimate to update the technology for your liquefied natural gas pipeline, which was constructed ten years ago. Now, you are working to develop you budget. But you have some funding limits to consider as you do so, which means you—

   a. Follow your financial framework
   b. Have imposed date constraints
   c. Acquire additional funds from outside sources
   d. Use retained earnings to cover short-term gaps

40. Your company, noted for its use of innovative technology in its work, is also noted for exceeding its budget. On your project, you have been asked by the executive team to keep costs under control. You are focusing your attention on—

   a. Informing your stakeholders of approved change requests and their costs
   b. Using earned value and following the 50/50 rule
   c. Focusing on work performance information as you begin your work to control your project costs
   d. Involving stakeholders to ensure change requests are decided as quickly as possible

41. While managing a large project in your organization, you realize that your project team requires training in contract administration because you will be awarding several major contracts. After you analyze your project requirements and assess the expertise of your team members, you decide that your team will need a one-week class in contract administration. This training should—

    a. Commence following guidelines in the resource management plan
    b. Commence as scheduled and stated as part of the procurement management plan
    c. Be scheduled if necessary after performance assessments are prepared and after each team member has had an opportunity to serve in the contract administrator role
    d. Commence as scheduled and stated in the team development plan

42. Assume that on your project, you are using earned value management. You also are conducting performance reviews. You recognize, though, that an important aspect of cost control is determining the cause and variance compared to the cost baseline. You have found that—

    a. The most frequently analyzed measurements are CPI and SPI
    b. The milestone method is the most accurate way to assess performance using earned value
    c. Graphical techniques will enhance how the information is received
    d. The range of acceptable variances will tend to decrease over time

43. Your project sponsor has asked you, "What do we now expect the total job to cost?" Given that you are using earned value, you should calculate the—

    a. To-complete performance index
    b. Estimate to complete
    c. Estimate at completion
    d. Budget at completion

44. One key reason that the Develop Project Charter process is so important is that it—

    a. Documents the boundaries of the project
    b. States the methods for acceptance of the project's deliverables
    c. Describes the project's characteristics
    d. Links the project to the organization's strategic objectives

45. Your company has been awarded a contract for project management consulting services for a major government agency. You were a member of the proposal writing team, are PMP® certified, and you are the project manager. You are now working to prepare your project management plan, which is to be submitted in one week. You decided to use some interpersonal and team skills to help develop your plan. While a number are possible, you selected—

 a. Conflict management
 b. Active listening
 c. Networking
 d. Political awareness

46. Assume you had a phase gate meeting with your Governance Board for your project to develop the next generation radar system as part of the nation's airspace modernization program. At this meeting, the Board approved your project management plan. However, as you begin to execute your plan, an organizational process asset to consider is—

 a. Sustainability
 b. Cost and budget management
 c. Legislation and regulations
 d. Performance management data base

47. Consider the data in the table below. Assume that your project consists only of these three activities. Your estimate at completion is $4,400.00. This means you are calculating your EAC by using which of the following formulas?

Activity	% Complete	PV	EV	AC
A	100	2,000	2,000	2,200
B	50	1,000	500	700
C	0	1,000	0	0

 a. $EAC = AC/EV \times BAC$
 b. $EAC = AC/EV \times$ [work completed and in progress] + [actual (or revised) cost of work packages that have not started]
 c. $EAC = AC +$ Bottom-up ETC
 d. $EAC = \%$ complete $\times$ BAC

48. Assume you are preparing your stakeholder engagement plan. As you do so, you want to make sure in the decisions you make that you—

    a. Hold meetings with your stakeholders
    b. Organize as much information as you can find about each stakeholder
    c. Use prioritization/ranking
    d. Use voting based on multi-criteria analysis

49. The lessons learned documentation is an output from the—

    a. Identify Stakeholders process
    b. Develop Project Management Plan process
    c. Manage Communications process
    d. Plan Communications Management process

50. Your experience has taught you that inappropriate responses to cost variances can produce quality or schedule problems or unacceptable project risk. When leading a team meeting to discuss the importance of cost control, you note that cost control is concerned with—

    a. Influencing the factors that create change to the authorized cost baseline
    b. Developing an approximation of the costs of the resources needed to complete the project
    c. Allocating the overall cost estimate to individual work items
    d. Establishing a cost performance baseline

51. You are pleased to be the project manager for a new video conferencing system for your global organization. You want it to be one that is easy to use and is state of the art. As the project manager, you also are the project leader. You want to create an environment to facilitate team work and motivate the team with challenges and opportunities. As you do so, you want to concentrate on—

    a. Agreement and trust
    b. Political and cultural awareness
    c. Negotiation and influencing
    d. Decision making

52. You are developing the next generation of your financial management system. In preparing your project management plan, you decided that you would use an agile approach rather than the typical waterfall approach your organization has used in the past. Agile is—

    a. Easier to control as the team is co-located
    b. An enterprise environmental factor
    c. An organizational process asset
    d. Useful in obtaining the views of those on the project team

53. Your company is in the project management training business. In addition, the company publishes several exam study aids for the PMP® and CAPM® exam. You have your PMP®, and you have been appointed as the project manager to make sure your company's training materials are updated to be aligned with the new *PMBOK® Guide*. You must complete your project in six months. Your schedule is complete, and you are working now to monitor and control it. Your company is testing agile on your project to consider whether all of its projects should use it. You need to be concerned about—

    a. Revising your schedule baseline
    b. Reprioritizing the remaining work plan
    c. Adjusting leads and lags
    d. Using modeling techniques

54. You are trying to determine whether or not to conduct 100% final system tests of 500 ground-based radar units at the factory. The historical radar field failure rate is 4%; the cost to test each unit in the factory is $10,000; the cost to reassemble each passed unit after the factory test is $2,000; the cost to repair and reassemble each failed unit after factory test is $23,000; and the cost to repair and reinstall each failed unit in the field is $350,000. Using decision tree analysis, what is the expected value if you decide to conduct these tests?

    a. $5.5 million
    b. $5.96 million
    c. $6.42 million
    d. $7 million

55. Motivation is dynamic and complex. The overall success of the project depends on the team's commitment to it, which is directly related to motivation. On your project, you want to create an environment to meet project objectives, but you want to motivate your team to—

    a. Recognize why decisions are made
    b. Encourage independent work
    c. Provide a high quality of information exchange
    d. Promote job satisfaction

56. Each time you meet with your project sponsor, she emphasizes the need for cost control. To address her concerns, you should provide—

    a. Work performance information
    b. Cost baseline updates
    c. Resource productivity analyses
    d. Trend analysis statistics

57. One output of the Control Costs process is cost forecasts, which is when—

    a. Modifications are made to the cost information used to manage the project and are communicated to stakeholders
    b. Trend analyses are performed and communicated to stakeholders
    c. A budget update is required and communicated to all stakeholders
    d. A calculated EAC value or a bottom-up EAC value is documented and communicated to stakeholders

58. You work for an electrical utility company and will be managing a project to build a new substation that will serve a new industrial park. This project was authorized because of a business need, and you have a defined business case for it. Not to be overlooked in the early stage of this project is the—

    a. Benefits management plan
    b. Options to address the business problem
    c. Identification of critical success factors
    d. Decision criteria for various courses of action

59. You are about to close your project. In doing so you are using trend analysis, which is—

    a. A form of data analysis
    b. A way to show corrective actions taken on the project
    c. A method of assessing the usefulness of causal analysis
    d. An approach that also involves expert judgment

60. In order to manage risk appropriately on your project, you want to know the thresholds of your key stakeholders for risk, which is—

    a. Stated in the risk register
    b. The risk appetite
    c. Explained in the risk report
    d. Used to prepare the RBS

61. You are following a collaborative approach as you lead you team and are conducting a team assessment in addition to individual assessments. The performance of a successful team is measured in terms of—

    a. Project commitment
    b. Achieving project objectives
    c. Enhanced ability to work as a team
    d. Effective decision making

62. Your project is considered very risky. You plan to perform numerous what-if scenarios on your schedule using simulation software that will define each schedule activity and calculate a range of possible durations for each activity. The simulation then will use the collected data from each activity to calculate a distribution curve (or range) for the possible outcomes of the total project. Your planned approach is an example of which of the following techniques?

    a. PERT
    b. Monte Carlo analysis
    c. Linear programming
    d. Concurrent engineering

63. In Monitor Communications, you want to ensure the communications needs of stakeholders are met. It then is useful to—

    a. Review the communications strategy
    b. Negotiate with stakeholders
    c. Conduct ad hoc meetings
    d. Establish a knowledge management repository

64. You and your team are preparing your program management plan. While much of your plan reflects plans developed in other processes, one plan that often is overlooked is the—

    a. Benefit realization plan
    b. Knowledge management plan
    c. Staffing management plan
    d. Configuration management plan

65. Lessons learned are important throughout project management. In fact, your lessons-learned register is updated as an output of the manage stakeholders engagement process, so it contains—

    a. Root causes of issues
    b. Effective conflict management approaches
    c. Effective and ineffective approaches
    d. Approved changes

66. Each project is unique. Assume you are managing a project and are in the process of developing your schedule for it. However, your Enterprise Project Management Office has mandated a detailed methodology that must be followed on all projects. You have asked for a waiver as you are managing a project in which you need specific subject matter experts available at specific times. In the Estimate Activity Duration process, you need to consider—

   a. Adding more lags
   b. Doubling the number of needed resources
   c. Use of resource leveling
   d. Use of resource smoothing

67. As stakeholders engage with the project, your goal is to monitor stakeholder relationships and tailor strategies by modifying engagement strategies and plans. As the project manager, this means you should—

   a. Revise the stakeholder matrix
   b. Review the roles of the stakeholders
   c. Update the project management plan
   d. Update the stakeholder engagement plan

68. Once you were appointed project manager and worked to prepare your charter and project management plan, you realized monitoring and controlling transcended the project. You realized change is common on projects, and you wanted to exploit and embrace it. This meant you—

   a. Required configuration management
   b. Updated the cost baseline
   c. Updated the project management plan
   d. Could use a project management information system (PMIS)

69. Assume you are new to your company and you are working as a project manager to make your IT department more customer centric. You worked with you team and identified the key stakeholders. Now, you are preparing your stakeholder engagement plan. You realize additional information about the environment in which the company works would be helpful, so you decide to—

   a. Review the organizational process assets
   b. Consult the communications management plan
   c. Ask some experts
   d. Use your supportive stakeholders from the stakeholder engagement matrix

70. Assume your project is to last three years. You know you will have a lot of stakeholders who are interested in just one phase of your project, and also during this time, you expect some team members will move to advance their careers, and new team members will join the project. As you prepare your project management plan—

    a. Once it is complete, you next will baseline it
    b. It needs to be robust
    c. You need to ensure it has a stakeholder threshold
    d. It needs to be comprehensive since it is prepared one time

71. Activity attributes are used to extend the description of the activity and to identify its multiple components. In the early stages of the project, an example of an activity attribute is—

    a. Activity classification
    b. Activity description
    c. Predecessor and successor activities
    d. Activity name

72. You are working on a new project in your city to construct an environmentally friendly landfill. The existing site is so undesirable that many residents have moved to other neighboring cities because of their proximity to it. However, even though the project has the support of the public, you need to have a number of hearings with the city's government before you are authorized to begin work. As you are in the planning phase of the project, you are waiting for these hearings to be scheduled and held before you can begin site preparation. These hearings are an example of—

    a. A milestone
    b. An external dependency
    c. An item to be scheduled as a fragnet
    d. A mandatory dependency

73. You are working on a project and want to know how many activities in the previous month were completed with significant variances. You should use a(n)—

    a. Control chart
    b. Inspection
    c. Scatter diagram
    d. Trend analysis

74. During your project, designed to last three years, to produce flying automobiles in certain parts of your country to avoid extreme traffic congestion, you know team members will not be on your project the entire time. Therefore, your goal with your team members and your contractors is to focus on managing project knowledge. You want to avoid a misperception, which is—

    a. Using tangible deliverables
    b. Creating an atmosphere of trust
    c. Focusing on codified explicit knowledge
    d. Realizing it is essential to have a co-located team

75. Two of your team members, who are subject matter expert in Cloud computing, are having a technical conflict. While you lack their detailed level of expertise in this area, you do know enough about it to discuss their concerns and hopefully resolve the conflict they are having. Your goal is to harmonize relationships meaning you are using which conflict resolution approach?

    a. Avoiding
    b. Accommodating
    c. Compromising
    d. Collaborating

76. Assume after working for a year with a team of four people to assist you, your project to convert all of the organizational process assets in the company to ones that can be easily accessed electronically is complete. You are now closing your project. As you do so, you realize many project documents will be updated. Of particular importance is the—

    a. Assumption log
    b. Lessons learned register
    c. Issue log
    d. Risk register

77. You are working on a mega construction project located in the middle of a jungle. Much of the work is outsourced. Your company is known for being a leader in the field and also in project management. You are using the Building Information Model. This means you—

    a. Can award contracts quickly
    b. Are able to manage the Conduct Procurements process more effectively
    c. Can reduce construction claims
    d. Will not require warranties

78. You are beginning a new project staffed with a virtual team located across five different countries. To help limit conflict and misunderstandings concerning the justification, objectives, and high-level requirements of the project among your team members and their functional managers, you ask the project sponsor to prepare a—

    a. Memo to team members informing them that they work for you now
    b. Project charter
    c. Memo to functional managers informing them that you have authority to direct their employees
    d. Human resource management plan

79. You want to make sure that when you complete your project to merge two large government agencies in your country that everyone will be pleased with your overall result. While this project is a major undertaking, and many people fear they will lose their jobs in the process, your goal is to have a quality project once you and your team finish the work. Quality is everyone's responsibility. To best Control Quality, you need to—

    a. Detect and control any defects or issues before you finish the project
    b. Maintain engagement with all stakeholders during the project
    c. Start by preparing system or process flowcharts and then use them
    d. Ensure the culture of the two organizations is committed

80. Many different factors can influence the project team. As the project manager and leader, one should be aware of

    a. Staffing management methods
    b. Options broadening
    c. Organizational change management
    d. Multiplying options by shuttling between the specific and the general

81. Recently, your company introduced a new processing system for its products. When you did so, you outsourced 75% of the work as you and your team focused on involving the internal stakeholders in the process and preparing them for the changes. To handle the outsourcing successfully, you need to—

    a. Use inspections
    b. Focus on short-term gains, so stakeholders can see value
    c. Create value for your team and for the suppliers
    d. Ensure supplier conform to requirements

82. As the project manager, you are negotiating with functional managers and other project managers to staff your project with the required levels of expertise. To determine the most appropriate criteria to select your team, you decided to—

    a. Use expert judgment
    b. Establish a virtual team
    c. Set clear expectations
    d. Use multi-criteria decision analysis

83. On your manufacturing project, management realizes that immediate corrective action is required to the material requirements planning (MRP) system to minimize rework. You lack the needed resources and costs are now overrunning the budget. To implement the necessary changes, you should follow—

    a. The resource acquisition policy
    b. The resource management plan
    c. Agreements
    d. A defined integrated change control process

84. You are the project manager on a project to improve traffic flow in the company's parking garage. Increasingly a lot of accidents have occurred as people are in a hurry, and often there has been road rage. Your plan is to achieve customer satisfaction with your new approach. You do not want problems after your work is complete and you decide to—

    a. Help anticipate how problems occur
    b. Use questionnaires and surveys
    c. Show the results of your process
    d. Forecast future outcomes

85. You are working on an international project in which some of your 25 team members will be working in a country in which your company lacks previous experience. As you work to negotiate for people to work on your project, since you work in a matrix structure, you realize you need to—

    a. Provide special consideration to external policies
    b. Assess the culture of this new country
    c. Review personnel administration policies
    d. Focus on communications planning

86. Schedule control is one important way to avoid delays. While planning and executing schedule recovery, one tool available to you for Control Schedule is—

    a. Changing the schedule management plan
    b. Immediately rebaselining
    c. Adjusting leads and lags
    d. Holding performance reviews

87. You have been the project manager for your nuclear submarine project for four years. While you did not assume this position until the project management plan had been prepared and approved, you find you spend a significant amount of time collecting data and communicating. You also spend time reviewing the impact of project changes and implementing ones that have been approved. Often you have had to modify a non-conforming product, which means you are spending time on—

    a. Corrective actions
    b. Updating the project's requirements
    c. Updating the traceability matrix
    d. Defect repair

88. You were assigned recently as the project manager of a program management office project to implement a new enterprise-wide scheduling system for use throughout your company. You identify the need for a project charter to provide you with appropriate authority for applying resources, completing the project work, and formally initiating the project. Who should issue the project charter?

    a. The project manager—you
    b. The customer
    c. Someone external to the project
    d. A member of the training and development department as they will own the training on the new system

89. Your customer has expressed an interest in becoming more involved on your project, and you believe it would be desirable, so the customer and his representatives get to know you and your team better and can establish a great working relationship. You have decided to—

    a. Include your customer as you gather requirements
    b. Provide your customer with copies of the plans for your project before the governance board approves them
    c. Invite your customer and his team to project meetings
    d. Use a social media approach with a common platform

90. You are managing a construction project located in a desert to build a new way to generate electrical power. It is in this remote location as stakeholders will not know about it and become negative toward it if it were near their homes. To ensure physical resources arrive as specified, the project manager must—

    a. Avoid a large expenditure of resources
    b. Follow the project schedule
    c. Specify in detail the problem and use problem solving to solve it
    d. Identify and deal with resource shortages

91. You are working to control resource use on your manufacturing project, which requires different types of physical resources. You want to allocate and use the necessary materials in a timely and effective way. Fortunately, your organization specializes in this area and has a knowledge repository available, which is easy to access and use. It includes—

    a. Process analysis
    b. Industry-specific resources
    c. Resource configurations
    d. Total productive maintenance

92. You are preparing your cost estimate for your project in robotics. Already, some competitors have learned about this product, and they want to be possible purchasers of the project when it is complete. You consider this to be an opportunity, which means—

    a. Your cost estimate's accuracy rate should be from −5% to +10%
    b. You need to reduce activity costs as much as possible
    c. Your estimate should include the cost of financing
    d. You should include indirect costs in your estimate

93. You are planning a project and want to account for how the project will be managed in the future. While building your cost performance data, you want to provide guidance for when the project is later executed, because you know that different responses are required depending upon the degree of variance from the baseline. For example, a variance of 10 percent might not require immediate action, whereas a variance of 20 percent will require more immediate action and investigation. You decide to include the details of how to manage the cost variances in the—

    a. Cost management plan
    b. Change management plan
    c. Performance measurement plan
    d. Variance management plan

94. Assume that you are managing a project team. Your team is one in which its members confront issues rather than people, establish procedures collectively, and is team oriented. As the project manager, which of the following represents your team's stage of development?

    a. Storming
    b. Norming
    c. Adjourning
    d. Performing

95. You are finalizing your project and after two years, you are ready to close it. Before doing so, you need to—

    a. Manage knowledge transfer
    b. Ensue you met your planned success criteria
    c. Conduct a review with your three contractors
    d. Resolve all risks

96. You are working on a project and want to identify the cause of problems you are having as you Control Scope. You wonder if you need to tailor some of the processes in your organization to better meet your project's ability to be a success. You realize in the area of Control Scope it is not necessary as—

    a. More inspections can be done
    b. Informal scope control related practices are used
    c. You have work performance information to review
    d. You can use walkthroughs

97. You are working on a construction project in a city different from your headquarters' location. You and your team have not worked in this city, City B, previously, and you lack knowledge of the local building codes. You had a team member review the codes, and he said they were in far greater detail than those in your city, City A. In addition, your new project in City B is to take three years to complete and will use some unproven technology. Plus, your company wants you to follow agile, so you decided to collect requirements in user stories. In this situation, your best scheduling approach is—

    a. Agile release planning
    b. On-demand scheduling
    c. Iterative scheduling with a backlog
    d. Critical chain

98. Assume that you are managing a project that once completed will take your company into new markets. Since it is so significant, it has the interest of executive managers and other key internal stakeholders. You know for success, the marketing department will play a key role. You and your team identified them as key stakeholders. You met with the Chief Marketing Officer, and she indicated she would support the project. However, each time you have had a status review meeting with the executive group, except for the first meeting, the Chief Marketing Officer has not attended subsequent meetings. She also has not sent someone from her staff. This situation shows the importance of—

    a. Executive support
    b. The need to escalate this issue to your sponsor
    c. Maintaining the stakeholder register
    d. Engaging stakeholders at certain stages

99. The nature of project work is such that it inevitably changes. You know this is the case on your software project as you were about 50% done when the company announced all software work was to be done using agile, even work in progress, and your project was using waterfall. You now believe it is time to ensure your stakeholders have not changed, and you want to monitor and assess stakeholder engagement levels. You decide to—

    a. Review work performance information
    b. Hold some meetings
    c. Use subject matter experts
    d. Reevaluate your power/interest grid

100. Communications develops the relationships for successful project outcomes. They can vary from short e-mails to formal meetings and presentations. There are two parts to successful communications. The first part is—

    a. Preparing the communications management plan
    b. Determining whether stakeholders are internal or external
    c. Assessing cultural differences
    d. Developing a communications strategy

101. You are in the early stages of a project to manufacture disposable medical devices. You are preparing your schedule management plan for your project. As you do so, you plan to use—

    a. Alternative analysis
    b. The business case
    c. Different types of dependencies
    d. Your project management information system

102. You are managing a two-year project to transform your company, so it is more customer facing. No longer will work be done in silos, and instead, cross-functional teams will be used. The goal is the customer will know exactly whom to contact in the company with a question and will not have to make a number of calls to find the right person. Since this approach is one of the company's strategic priorities, the project management plan—

    a. Requires subsidiary plans from the other nine knowledge areas to be part of the plan
    b. Relies on organizational assets to complete it
    c. Needs stakeholder buy it at all levels to ensure success
    d. Should be consistent with the portfolio management plan

103. You are managing a two-year project to transform your company, so it is more customer facing. No longer will work be done in silos, and instead, cross-functional teams will be used. You are preparing your communications management plan. It should contain—

    a. Stakeholder directory
    b. Resources for communications activities
    c. Updates to the project management plan
    d. Updates to the project schedule

104. Obviously as a project manager, you will be making decisions throughout the project. A guideline for decision making is—

    a. Active listening skills to ensure all points of view are heard
    b. Focus on the vision of the project
    c. Gather relevant or critical information
    d. Focus on goals to be served

105. You and your sponsor are working on your project charter. It is important since you are working a totally functional organization, but for this project to be considered a success, you need people on your team that work in these functional departments. To make sure the functional managers will sign off on the charter and commit the resources you need at the time you need them, you decide to—

    a. Conduct interviews
    b. Have meetings
    c. Convene a focus group
    d. Bring in a facilitator and conduct a brainstorming session

106. Your company has changed to an agile environment, and you are managing your first project using agile methods. After obtaining your certification in agile, you know through your studies of it plus the training your company provided that an effective way to Monitor Risks is to—

    a. Review the overall risk effectiveness of the project
    b. Follow the requirements management plan
    c. Have fallback plans ready
    d. Accelerate knowledge sharing

107. You are the project manager for a major logistics installation project and must obtain specific services from local sources external to your project. You are working now to prepare your project charter and decided to consult with some experts in—

    a. Benefits realization
    b. Contracts and procurement
    c. Organizational strategy
    d. Assumptions analysis

108. Assume you are finally in the Close Project phase after almost two years on your project. You are holding a series of meetings, one of which is—

    a. Configuration management
    b. Reporting
    c. Requirements traceability review
    d. Contractor assessments for the qualified seller list

109. You are working on a project to upgrade the existing fiber-optic cables in your province. You have identified your stakeholders and now are preparing your stakeholder engagement plan. To assist you as you do so, you know your managers prefer visual representation of any possible information on any topic, so you decide to use—

    a. Stakeholder efficiency factors
    b. Mind mapping
    c. The leaders from your stakeholder engagement matrix
    d. Root-cause analysis

110. You are working extremely hard to manage the engagement of your stakeholders. You hold extensive meetings with those whom you have classified as leading, so they can assist you in changing resistors to be at least neutral. You also are using a variety of forums to connect with them and to continually identify new stakeholders. You decided to adapt agile since it is a software project and hold sprint planning meetings plus daily stand-up meetings with your team. Lately, even though the team had been working well together, and you consider it to be in the performing stage for whatever reason, they now are storming, and team members are becoming resistant stakeholders. You need to—

a. Use conflict resolution skills
b. Meet with each person and find out if the person still wants to be on the team
c. Refer to the team charter
d. Use active listening

111. You have had five contractors working for you on your legacy systems conversion project, which has taken two years. Finally, you have finished your work, users seem pleased with your results, and you are closing your project. However, before doing so, you need to close your contracts. This means you need to consider—

a. Statement of work in the contract
b. Technical documentation
c. Product description
d. Organizational process assets

112. After two years, you developed a new approach to mentor people before they joined the company to determine if they would enjoy working there, and then if they were hired, you assigned a mentor to work with them. People seem pleased with the results of your work, and you now need to—

a. Reassign staff
b. Update your records
c. Transfer your work to human resources
d. Involve stakeholders

113. You decided to implement a team-based reward and recognition for your project, and the team members agreed in the project's ground rules that were established. However, while your sponsor liked the idea, the human resources department states individual performance appraisals must be done. It is now time to close the project. As the project manager, you decide to—

    a. Evaluate performance when a team member completes work on his or her deliverable or activities
    b. Release resources when the project is in the closing phase
    c. Follow the person's Individual Development Plan
    d. Talk with the person's functional manager

114. Your company is embarking on a project to launch a new product delivery service. You are the project manager for this project, and you have developed your charter working with your sponsor and some other key stakeholders. As you prepared your charter, you realized you also should prepare a(an)—

    a. Summary milestone schedule
    b. List of overall risks
    c. Assumption log
    d. Assessment of financial resources required

115. Assume you are working on a software project to implement a new warehouse management system. Since your team is collocated, your sponsor felt it was a perfect project to test whether agile methods are appropriate for the company. As you work to do this project, you realize it is necessary to—

    a. Have high emotional intelligence
    b. Emphasize collaboration
    c. Use key performance indicators
    d. Have regular communications

116. You are preparing your project resource plan. In it, you will use some chart techniques to help determine resource requirements. As the project manager, you know it is your responsibility to spend time in acquiring, managing, motivating, and improving your team. You also know that—

    a. A kickoff meeting is recommended.
    b. Involving team members in planning is beneficial.
    c. If ground rules are set, the number of conflicts and changes will be lessened.
    d. Rewards and recognition will be handled smoothly throughout the project.

117. Team building should be ongoing throughout the project life cycle. However, it is hard to maintain momentum and morale, especially on large, complex projects that span several years. One guideline to follow to promote team building is to—

    a. Use it to build a collaborative working environment
    b. Conduct team building at specific times during the project through off-site meetings
    c. Engage the services of a full-time facilitator
    d. Develop the project schedule with times set aside for team building

118. You have been assigned as the project manager for a major project in your company where the customer and key supplier are located in another country. You have been working on your project for six months. Recently, you traveled to this country, and at the conclusion of a critical design review meeting, which was highly successful, you realized you were successful in building a high-performing team. You had your own team members, who work in a weak matrix structure, on a conference call during this meeting. Although it was difficult to reach agreement on some key issues, you therefore relied on your interpersonal skills in—

    a. Facilitation
    b. Conflict resolution
    c. Influencing
    d. Decision making

119. You feel fortunate to be assigned as the project manager on a multi-phase project that was requested by your company's key customer. It has the interest of the senior executives, and it was approved by the organization's portfolio oversight group. As you begin to work on this multi-phase project, a best practice is to—

    a. Periodically review the business case
    b. Establish a Governance Board to conduct the phase reviews
    c. Use project audits
    d. Focus on realizing benefits

120. Because risk management is relatively new on projects in your company, you decide to examine and document the effectiveness of the risk management process. You therefore—

    a. Conduct a risk audit
    b. Hold a risk status meeting
    c. Ensure that risk is an agenda item at regularly scheduled staff meetings
    d. Reassess identified risks on a periodic basis

121. Thinking back to lessons that your company learned from experiences with its legacy information systems during the Y2K dilemma, you finally convinced management to consider Software as a Service. You worked with your sponsor as you prepared your business case, and it was approved. Now you are working to develop your project charter. To ensure you are able to attain the key resources you need at the they are needed you recognize—

    a. You need to communicate actively with the functional managers
    b. You first need a verified stakeholder list
    c. Conflict management skills are useful
    d. Expert judgment should not be forgotten

122. On your systems development project, you noted during a review that the system had less functionality than planned at the critical design review. For example, too many defects were identified. This finding suggests that during the Monitor Risks process you should use which following tools and techniques?

    a. Risk reassessment
    b. Variance analysis
    c. Technical performance analysis
    d. Reserve analysis

123. You have found as you Monitor Risks that some risks have not been closed, and some secondary risks have occurred but have not been documented. You decide you should—

    a. Issue a change request
    b. Conduct an audit
    c. Hold a meeting
    d. Review work performance information

124. Assume you are working on a mega project to upgrade every bridge in your state. You have outsourced most of the work given the estimated bridges of varying sizes approaches 3,000. You have 250 contractors, and many of them have some subcontractors. In this situation as the buyer, you realize—

    a. An effective contact type is the fixed-price-incentive
    b. You may accept more risks
    c. You can transfer risks to the contractors
    d. You need a team member who can be dedicated to resolving claims

125. A structured review of the seller's progress to deliver project scope and quality within cost and schedule is known as a(n)—

    a. Procurement performance review
    b. Procurement audit
    c. Inspection
    d. Status review meeting

126. Within your company's portfolio, your project is ranked in the top five in terms of importance of the 60 projects under way. You are working in your construction company and now are preparing your quality management plan. While quality is critical to success, you are determining the cost of quality and have divided the costs into the costs of conformance and those of nonconformance. An example of a cost of nonconformance is—

    a. Destructive testing loss
    b. Warranty work
    c. Time to do it right
    d. Inspections

127. Assume you are developing your project charter and decided to review enterprise environmental factors. An example is—

    a. Workmanship standards
    b. Organizational processes
    c. Monitoring and reporting methods
    d. Results of previous project selection decisions

128. Each project is unique, and not all methodologies will fit every team. Assume you are managing your project to develop the world's largest wooden roller-coaster as a tourist attraction for your city. You are working to develop your team, so it is in the performing stage. As you review how resource management policies are used in your company, you realize you need to consider—

    a. Matrix management
    b. Diversity
    c. Resource management methods
    d. Required skills

129. On your project as you prepare your stakeholder engagement plan, you decided it would be worthwhile to prepare a stakeholder engagement assessment matrix. You realized by doing so, you could—

    a. Update the stakeholder register
    b. Consult with experts
    c. Assigning a team member to work with stakeholders in each category
    d. Direct the right level of communications to the stakeholders

130. Assume you are identifying your stakeholders. You want to use a three-dimensional model, so you decide to—

    a. Combine the power/influence grid with the power/influence grid
    b. Use a salience model
    c. Use a stakeholder cube
    d. Establish three directions of influence

131. Work completed, key performance indicators, technical performance measures, start and finish dates of schedule activities, number of change requests, number of defects, actual costs, and actual decisions are examples of work performance data, which are an output of—

    a. Project Plan Development
    b. Risk Control
    c. Monitor and Control Project Work
    d. Direct and Manage Project Work

132. Two team members on your current construction project are engaged in a major argument concerning the selection of project management software. They refuse to listen to each other. However, this conflict now is causing schedule delays. The most appropriate conflict resolution approach for you to use in this situation is—

    a. Accommodating
    b. Compromising
    c. Collaborating
    d. Forcing

133. You are preparing your project management plan to ensure the safety of camel milk in your country. As you prepare your project management plan, you want to ensure that the starting point for it is—

    a. The business case
    b. The project charter
    c. Organizational process assets
    d. The selected project life cycle

134. A key member of your project has deep technical skills and many years of experience in the company. However, she felt she should be the project manager. When you became the project manager instead, her morale deteriorated. You worked with her to obtain her ideas whenever there were issues and commended her work to others. But her morale is so low, and she is constantly complaining. Now you notice her morale is so poor that it is affecting other team members, and there are numerous negative conflicts you need to resolve. You have decided you need to reassign this staff member and have worked with your PMO manager to do so. Your next step is to—

   a. Meet with this team member and inform her of your decision
   b. Inform the team she is moving to a new position but recognize her contributions to them
   c. Issue a change request
   d. Work with human resources and then meet with this team member

135. You are performing a stakeholder analysis on your project, and you are working with your team to ensure you do not omit a key stakeholder. Your objective in doing this analysis is to—

   a. Prepare a stakeholder register
   b. Determine the model you plan to use to classify the stakeholders
   c. Assess how certain stakeholders are likely to respond in certain situations
   d. Identify all potential stakeholders and their relevant information

136. Before considering a project closed, what document should be reviewed to ensure that project scope has been satisfied?

   a. Project scope statement
   b. Project management plan
   c. Project closeout checklists
   d. Scope management plan

137. You have just been promoted to be a mega project manager in your company. The project will be using outsourcing, and some agreements are in place and signed. Your project is in three different locations. You want to be effective in controlling the procurements on your project. You decided to—

   a. Hold daily meetings with each seller
   b. Acquire a systems integration contractor
   c. Follow an adaptive approach
   d. Determine the variance thresholds to be used in the project

138. Although you have closed your project, you still have some outputs even though your deliverables have been met, the administrative closure tasks have been completed, and the project plan has met its objectives. You now need to—

    a. Validate the models used for future projects through variance analysis
    b. Improve the metrics used through trend analysis
    c. Assess the schedule objectives to see if benefits were achieved
    d. Analyze the interrelationships for future projects

139. You have a conflict on your team but have enough time to resolve it, and you want to maintain future relationships. Thankfully, there is mutual trust, respect, and confidence among the parties involved. You decide to use collaborating to resolve this conflict. In using this approach, your first step should be to—

    a. Separate people from the problem
    b. Identify the causes of the conflict
    c. Establish ground rules
    d. Explore alternatives

140. You are working hard to make sure the guidelines in your communications management plan are not only followed but are used. You realize language is a major factor in communications activities, especially with global teams or teams that are comprised of people from different cultural groups. A first step in this process in Monitor Communications in this area is to—

    a. Hold a meeting
    b. Review the stakeholder register
    c. Determine the number of languages used
    d. Evaluate stakeholder involvement

141. Validate Scope works hand-in-hand with Control Quality and generally follows Control Quality. A tool and technique used in Validate Scope that is not used in Control Quality is—

    a. Decision making
    b. Inspection
    c. Statistical sampling
    d. Variance analysis

142. Assume your project is considered to be extremely important to your company as it is for its top client. You have been given the authority to assign resources to the project as you set up your team. An important criterion is—

a. Experience
b. Ability
c. Knowledge
d. Skills

143. You are a goal-oriented project manager who is more interested in work accomplishment than relationship building. This indicates that you tend to resolve conflicts primarily through the use of—

a. Smoothing
b. Compromising
c. Collaborating
d. Forcing

144. You are working on a long-term project that has a number of benefits to its customers and users. Therefore, as the project manager, one of your first steps was to identify the stakeholders that were critical to project success. You decided to use focus group sessions, which are—

a. Interviews with key stakeholders
b. Part of questionnaires and surveys
c. A cross-functional group
d. A form of the Delphi approach

145. There are many different techniques for effective communications. As you know communications represents about 90% of the project manager's time, you want to select the most appropriate one to use. You decide to focus on—

a. The choice of media
b. Mentoring
c. Communication styles assessment
d. Team building

146. Assume you and your team have prepared your risk management plan, identified, analyzed, and prepared your risk responses, and documented these data in your risk register. Now it is time to Implement Risk Responses. With all the planning you have done, it is time to manage the risks. You have decided that several agreed-upon risk responses were not implemented, and the risk occurred, but nothing was done. You realize the problem was the cost of implementing the response against the stakeholders' risk thresholds. Now you should—

    a. Review the risk thresholds with your sponsor
    b. Have each risk owner contact affected stakeholders about the risk they are managing
    c. Issue a change request
    d. Update assumptions and constraints

147. The key output of Identify Stakeholders that documents identification information, assessment information, and classification is the—

    a. Stakeholder management plan
    b. Communications plan
    c. Stakeholder register
    d. Communications log

148. An example of a tool and technique to create and connect people to information in Manage Project Knowledge is—

    a. Library services
    b. Storytelling
    c. Discussion forums
    d. Knowledge fairs

149. A key output from Estimate Costs—

    a. Cost baseline
    b. Cost of quality assumptions
    c. Reserve analysis
    d. Basis of estimates

150. A contract as a type of agreement typically is used when a project is being performed for an external customer. Agreements of all types are used as an input to—

    a. Develop Project Charter
    b. Develop Project Team
    c. Plan Procurement Management
    d. Conduct Procurements

151. As you prepare to close your project, which of the following is an input to the Close Project or Phase process?

    a. Work performance information
    b. Expert judgment
    c. Accepted deliverables
    d. Change requests

152. Maintaining the scope baseline is the main benefit of the process of Control Scope. As you work to avoid scope creep on you project, you want to focus on determining the cause and degree of difference from this baseline and project performance. To do so, you decide to—

    a. Establish scope guidelines
    b. Set up and follow a requirements traceability matrix
    c. Set up scope categories
    d. Use trend analysis

153. Assume you have been working with your sponsor to prepare you charter, and you plan to present it to your Steering Committee on Friday. You are managing a software project, and the business need stated that you should use agile for the first time in your company rather than waterfall. In the Develop Project Charter process, this is then—

    a. A tool and technique
    b. Part of the enterprise environment factors as an input to this process
    c. A high-level requirement
    d. Stated in the strategic plan as a tool and technique in this process

154. Procurement documents are used in the Identify Stakeholder process because they—

    a. Are an enterprise environmental factor and an input to the process
    b. Are an organizational process asset and an input to the process
    c. Note key stakeholders as parties in the contract
    d. Serve as a way to prioritize and classify stakeholders

155. You completed your stakeholder analysis. You also are working on a project that is small, and it is scheduled to last four months with a budget of $50,000 US. You have identified 30 stakeholders and now want to categorize them. You decide to—

    a. Use an impact/influence grid
    b. Prioritize them
    c. Use a salience model
    d. Determine their directions of interest

156. Working on a two-year project, you know you will face a number of problems, gaps, inconsistencies, and conflicts. Having been a project manager for the last five years, you know they will occur unexpectedly and may affect project performance so, you decide to—

    a. Use an assumption log
    b. Use an issue log
    c. Focus on effective use of change requests
    d. Realize you will need to update project documents

157. You are working on a project that needs approval from your City Council and the courts, because the project is one with significant environmental and social impacts. Although many consumer groups are advocates of this project, others are opposed to it. Hearings are scheduled to resolve these issues and to obtain the needed permits to proceed. In preparing your resource plan, you decide to designate a person as the court liaison, which is an example of a—

    a. Role
    b. Responsibility
    c. Required competency
    d. Ability of the team member to make appropriate decisions

158. Assume you are managing an international project. Your team is located in Atlanta, Georgia, US; Berlin, Germany; and Melbourne, Australia. You and your sponsor are located in Paris, France; and your customer is located in Athens, Greece. Recognizing the different locations of the stakeholders in your project in its initial stages, a best practice to follow in terms of working toward business value as you identify your stakeholders is to—

    a. Determine who decides the project is a success
    b. Determine how and when project benefits will be delivered
    c. Establish the project culture with a focus on excellent stakeholder relationships
    d. Identify basic cultural characteristics on your project

159. As a project manager, you recognize the importance of actively engaging key project stakeholders on a project. First however, as you identify your stakeholders, you should—

    a. Focus on relationships necessary to ensure success
    b. Review your project charter
    c. Review the communications management
    d. Focus on each stakeholder's power relevant to the project

160. You decided as you prepare your communications management plan for your project to reorganize your organization to document in it fundamental activities of effective communications activities and develop effective communication artifacts that you need to—

    a. Rely more on the stakeholder engagement matrix
    b. Include an issue log
    c. Understand as much as possible about the communications receiver
    d. Include communications data in the PMIS

161. As you prepare your project's quality management plan, you identified the various roles and responsibilities you need so your project achieves the customer's requirements, and they are satisfied with it. However, you work in a matrix organization, and you are concerned that even with your project charter, you may not have the needed resources available when you need them. Therefore, you realize—

    a. You need strong negotiation skills to work with the leaders of the various functional departments
    b. Management is responsible for providing needed resources
    c. You should design some experiments to see when you can use less experienced people and present your results to the management team
    d. You should work with your program sponsor to acquire the resources

162. You are managing a major international project that involves multiple teams in different locations. You are preparing your quality management plan for your project. The different people working in different geographic locations is a(n)—

    a. Constraint
    b. Easily overcome with your configuration management plan
    c. Organizational process asset
    d. Enterprise environmental factor

163. Infrastructure and engineering projects are increasing since the economy has been stagnant except for the past six months. Many multi-billion dollar projects are common, and you are leading one for your company. Many procurements are international. As the seller, you want to—

    a. Obtain a cost-plus-fixed-fee contract
    b. Avoid the need to compensate the buyer for buyer preparations
    c. Advertise in trade publications
    d. Work with the buyer for discounts

164. You realize risk management is a continuous process in project management as it is impossible to identify all risks. However, even though every member of your 12-person team has had training in risk management, some of the planned responses were not implemented. As the project manager, you are trying to determine why it was not done as planned. You decide to—

    a. Ensure your team knows the level of accepted risk appetite
    b. Conduct a cause-and-effect analysis
    c. Hold a meeting with your team to brainstorm ideas to fix the problem
    d. Set aside more budget as a contingency reserve

165. Assume you issued a Request or Proposal to update a shopping center in your city. Once you issued it, but before you awarded an agreement, a number of sellers asked questions. You decide to—

    a. Issue an amendment
    b. Capture all questions and provide written responses to them to the sellers that posed them
    c. Hold a bidder conference
    d. Withdraw the Request for Proposal and instead issue a Request for Information

166. Assume you are preparing your procurement management plan. One way to best manage and monitor the procurement is to—

    a. Have a meeting
    b. Use your risk register
    c. Review the requirements document
    d. Use your stakeholder register

167. Your project is considered to be one that is not predictable and has high variability. This is due to it being the first project in your aerospace company to venture into being a leader in space exploration and not just a bystander. Your firm is determined to lead the competition. In this environment as you plan needed resources for this project, you need—

    a. A collocated team
    b. Dedicated funding with a financial framework
    c. Agreements for fast supply
    d. Access to the people in the company with the competencies you need

168. Assume you are working on a complex project in your organization, which is in process of converting to agile methods. You are preparing your resource management plan and decided since agile is desired, a self-organizing team would be ideal. In such a team, you need to plan to—

    a. Use an adaptive project life cycle
    b. Focus on networking and professional development
    c. Have subject matter experts available when you need them
    d. Acquire generalized specialists

169. While you have worked as a project manager for ten years, you are new to your current organization. In the past, your organization was informal in terms of plans to prepare and processes to follow. Your new organization, however, requires that each project have a resource management plan. You are unsure of the primary effort level required to meet the project's objectives. You decide to—

    a. Hold meetings
    b. Use experts
    c. Develop a RACI chart and review it with your sponsor
    d. Use your WBS and develop a hierarchical organization chart

170. An intentional activity to ensure future performance of project work is aligned with the project management plan is—

    a. Preventive action
    b. Corrective action
    c. Implemented change requests
    d. Work performance information

171. There are numerous actions and activities as part of administrative closure to satisfy completion of exit criteria from a phase or when the project is closed. Assume on your project, you had four different contractors. To assist you as you close your project you—

    a. Review agreements
    b. Audit project success or failures
    c. Determine the contractors to be on the approved qualified vendor list
    d. Set up your lessons learned repository

172. As a project manager, not only must you be a leader, but you also must have outstanding skills in communicating. The importance of communications cannot be overstated even if you are working in an agile environment, which is what is used in your organization. Therefore, a key consideration is—

    a. Having an effective and efficient flow of communications between stakeholders
    b. Having an environment with an emphasis on transparent communications with your stakeholders
    c. Having resources available to ensure your stakeholders have access to needed information in a timely way
    d. Ensuring communications management policies, operating procedures, and processes are followed

173. Assume you are managing a project, and your project management plan has been approved. Your project has a high level of change associated with it. There is active and ongoing stakeholder involvement. This means you probably are working with a(n)—

    a. Adaptive life cycle
    b. Iterative life cycle
    c. Incremental life cycle
    d. Predictive life cycle

174. You are working to introduce formalized project management in your medical center, which is state-of-the-art. While it has done projects and has some under way, thus far they have been done in an ad hoc fashion. You were hired to move the center into a formal project management approach since you have your PMP, so you decided the best approach to demonstrate the value of project management was to use this culture change as a project. Now, you are preparing your project management plan. Because you know many people will resist this culture change, you decide to—

    a. Focus on the needed technical skills to develop the plan
    b. Gather data through focus groups
    c. Meet with people on project teams to see if they want to work with you
    d. First prepare a template for your plan

175. You are working to identify the stakeholders on your project. It is large and will last three years to develop a second location for your technology company. You know from academic research that—

    a. You need to determine your stakeholders' desired level of participation
    b. Contact information for each identified stakeholder is required
    c. You should perform an assessment to see how each stakeholder might react in certain situations
    d. Correct identification may lead to project success

176. Assume your project communication management plan has been approved by your sponsor and the members of your Steering Committee. You are managing a global project and have team members working virtually in four continents and stakeholders in numerous locations. Your next step is to—

    a. Set up an information management system
    b. Select communications technology
    c. Determine performance reporting methods
    d. Select a communications model

177. Although your project team is working virtually, you are striving to make it a high-performing team. You held a virtual kickoff meeting to ensure there was a shared project vision. You now see that team members are addressing the work to be done, but they do not seem to be collaborating. You realize the team is—

    a. Concerned about their formal roles and responsibilities
    b. Independent
    c. Forming
    d. Storming

178. Since your project is large and is to last three years to develop a second location for your technology company, you realize you have a large number of stakeholders, both positive and negative. You have completed your stakeholder analysis and have determined you have at least 155 stakeholders. Your next step is to categorize them, so you and your team spend your time later engaging with the key stakeholders that are critical to success. You decide to use—

    a. A stakeholder cube
    b. A power/influence grid
    c. Prioritization
    d. Directions of influence

179. Unknown unknowns seem to be more prevalent in environments and projects that are complex, seem to have many changes, and may be using destructive technologies. This means there is an increasing acceptance of these emergent risks leading to the need for project resilience on projects. Recognizing this may be the case on your complex project, you want to—

    a. Promote an integrated approach to project risk management
    b. Use warning lists
    c. Use influence diagrams
    d. Emphasize flexible processes

180. You realize that most people feel 'knowledge is power', it is hard to convince them that instead 'knowledge sharing is power'. Working on an innovative project that will last two years so your University can offer a PhD program in project management, you recognize you need to Manage Project Knowledge. As you work to do so, you should review the—

    a. Project team assignments
    b. Locations of team members
    c. Organizational, stakeholder, and customer culture
    d. Resource breakdown structure

181. You are working to put in a high-speed railroad line for passengers in a densely populated area in your country. Its goal is to be fast and fun to ride. It will connect three cities that now by car can take approximately six hours to drive to from the first point to the last point. Since it is new, you will be outsourcing much of the work. You got a number of responses from sellers for this work and using your source selection criteria, you have selected the one you think will provide the best product. You now want to—

    a. Establish a partnering agreement to reduce the possibility of a lot of claims and change orders
    b. Add risk management to the agreement
    c. Have a team member who will work with your procurement department and the seller on any key issues
    d. Set up earned value

182. One key interpersonal skill used to Manage Stakeholder Expectations is—

    a. Observation
    b. Building trust
    c. Compromise
    d. Networking

183. You are a project manager leading the construction project of a new garbage incinerator. Local residents and environmental groups are opposed to this project because of its environmental impact. Management agrees with your request to partner with a third party that will be responsible for providing state-of-the-art "air scrubbers," to clean the exhaust to an acceptable level. This decision will delay the project but will allow it to continue. It shows you are using which risk response strategy?

    a. Passive acceptance
    b. Active acceptance
    c. Mitigation
    d. Transference

184. After a year and a half, it is now time for you to close your construction project, and you had 15 contractors. First, you are working on closing procurements. For future lessons learned as you do so, you review the—

    a. Benefits realization plan
    b. 'As-built' plans
    c. Risk register
    d. Issue log

185. Working in the systems integration field, you are primarily responsible for coordinating the work of numerous contractors. Your current project is coming to an end, and you are closing it. You have 15 major contractors as well as a variety of other sellers. Now that you are closing contracts, you should—

    a. Conduct a trend analysis
    b. Use earned value to assess lessons learned
    c. Ask each contractor to meet with you individually at its own expense
    d. Review the business case

186. Most projects focus on risks that are uncertain and may or may not occur, which are risks such as a seller is no longer in business, requirements change after the design is complete, or contractors want to change existing processes. However, increasingly in project management many projects also have to identify non-event risks. Assume you must do so on your project. An example of a non-event risk is—

    a. The customer decides to terminate your contract
    b. A system test failure
    c. Inherent systems complexity
    d. One negative but powerful stakeholder does not attend meetings

187. You are identifying possible risks to your project concerning the development of a nutritional supplement. You want to aid your team in idea generation, so you decide to use—

    a. Documentation review
    b. Probability/impact analysis
    c. Checklist analysis
    d. Prompt lists

188. Managing five contractors on your project for a new stadium in your City that can be used for baseball and for football and can be easily converted for either sport is a challenge along with managing your 15-person project team. You have prepared your communications management plan for this process, and you feel your team is finding it useful. Given the nature of this project and the numerous stakeholders involved, you want to ensure the planned communications activities and documents have the desired effect. You want to make sure—

   a. The artifacts are being updated monthly
   b. The right message with the right content is delivered to the right audience
   c. If a project communications audit would be beneficial
   d. The risk register is updated

189. You are awarding another contract to serve as an integration contract on your stadium project. It has generated a lot of interest from potential sellers, and you now expect a number of proposals. Since you have never managed or awarded a systems integration contract before, you need to upgrade your interpersonal skills in—

   a. Leadership
   b. Negotiation
   c. Decision making
   d. Political awareness

190. In Monitor Risk, a burndown chart is an example of—

   a. Communicating whether the remaining reserve is adequate
   b. Displaying the number of resources involved in risk management
   c. Presenting the analysis of findings from a risk audit
   d. Showing the number of emergent risks and regularly identified risks to monitor

191. Your firm specializes in roller-coaster construction. It plans to bid on a soon to be released RFP to build the world's most "death-defying" roller coaster. You know that such a roller coaster has never been built before, and that this would be a high-risk project. As the buyer, you are determining your source selection criteria, You decide to use—

   a. Qualifications
   b. Quality and technical
   c. Sole source with a proven seller to you firm
   d. Quality and cost based

192. Work performance information in Control Scope can include—

    a. Recommended preventive or corrective action considered through change requests
    b. Categories of changes received
    c. Templates to the configuration management plan to see if they were used
    d. Ways to communicate scope changes

193. You want to ensure on your project that you focus on learning reviews before, during, and after the project. You are interested in the Manage Project Knowledge process as you believe your legacy conversion project will have a long-lasting impact on your company. These learning reviews are a(n)—

    a. Organizational process asset
    b. Project document
    c. Enterprise environmental factor
    d. Interpersonal skill

194. You recognize the importance of a lessons-learned register on any project. As you work to Manage Project Knowledge on your project to increase talent retention in your company, so there is less turnover and develop existing talent, a best practice to follow is to—

    a. Create it early in the project
    b. Recognize it is an output of the Manage Project Knowledge process
    c. Consider it as a standard organizational procedure as an organizational process asset
    d. Consider it as a deliverable

195. Products, services, and results in the marketplace are an input to which one of the following processes?

    a. Plan Procurement Management
    b. Conduct Procurements
    c. Control Procurements
    d. Close Procurements

196. Each project can benefit from stakeholder involvement; however, it is in both the project manager's and the teams' best interest to ensure that all project stakeholders have positive attitudes toward the project and its goals and objectives. Working as a project manager, you have a number of key stakeholders on your project. The stakeholder who provides funds and/or resources to the project is the—

    a. Owner
    b. Program manager
    c. Director of the project management office
    d. Contributor

197. Assume you are managing a contract for your project for a new generation of nuclear missiles. You realize since this project is complex, and will last four years, during this time, you will acquire a lot of information. You want to Manage Project Knowledge, and in doing so, you to decide to plan project communication based on the project and that of your company. This interpersonal skill is—

    a. Networking
    b. Facilitation
    c. Active listening
    d. Political awareness

198. Assume you are working to award an agreement as you develop an approach to make it easier to navigate all the various systems in the company to promote knowledge transfer. Before doing so, you should review the—

    a. Stakeholder management plan
    b. Bid documents
    c. Negotiation approach
    d. Requirements traceability matrix

199. Your project is being conducted in a series of phases, and reviews are held to determine if the project has met the criteria to go forward to the next phase. Your executives decided to adopt this approach to make it easier to cancel a project if it no longer added business value to the company and before it began to cost too much to continue it. You just learned today your project passed the concept phase, and next is the test phase. Throughout the project you have found your lessons-learned register to be especially useful as you work to Manage Project Knowledge. It includes, among other things, the impact, actions, and the proposed actions associated with various situations. As you move to the test phase, it means—

   a. You need to update your knowledge management plan
   b. The lessons-learned register becomes an organizational process asset
   c. The search terms to locate information in the lessons-learned register, or the meta data, require updates
   d. Safeguards on the intellectual property developed thus far need to be emphasized further

200. Assume you are working for a major airline in your country. Your project is to increase the timeliness in boarding flights even if people have a lot of carry-on luggage and also to install free Wi-Fi regardless of where one sits in the plane. You have identified your stakeholders and prepared your stakeholder register. In your stakeholder register, you should include—

   a. Assumptions
   b. Where the stakeholder has the most influence
   c. Stakeholder communications needs
   d. Stakeholder risks

Answer Sheet

1.	a	b	c	d	21.	a	b	c	d
2.	a	b	c	d	22.	a	b	c	d
3.	a	b	c	d	23.	a	b	c	d
4.	a	b	c	d	24.	a	b	c	d
5.	a	b	c	d	25.	a	b	c	d
6.	a	b	c	d	26.	a	b	c	d
7.	a	b	c	d	27.	a	b	c	d
8.	a	b	c	d	28.	a	b	c	d
9.	a	b	c	d	29.	a	b	c	d
10.	a	b	c	d	30.	a	b	c	d
11.	a	b	c	d	31.	a	b	c	d
12.	a	b	c	d	32.	a	b	c	d
13.	a	b	c	d	33.	a	b	c	d
14.	a	b	c	d	34.	a	b	c	d
15.	a	b	c	d	35.	a	b	c	d
16.	a	b	c	d	36.	a	b	c	d
17.	a	b	c	d	37.	a	b	c	d
18.	a	b	c	d	38.	a	b	c	d
19.	a	b	c	d	39.	a	b	c	d
20.	a	b	c	d	40.	a	b	c	d

41.	a	b	c	d		61.	a	b	c	d
42.	a	b	c	d		62.	a	b	c	d
43.	a	b	c	d		63.	a	b	c	d
44.	a	b	c	d		64.	a	b	c	d
45.	a	b	c	d		65.	a	b	c	d
46.	a	b	c	d		66.	a	b	c	d
47.	a	b	c	d		67.	a	b	c	d
48.	a	b	c	d		68.	a	b	c	d
49.	a	b	c	d		69.	a	b	c	d
50.	a	b	c	d		70.	a	b	c	d
51.	a	b	c	d		71.	a	b	c	d
52.	a	b	c	d		72.	a	b	c	d
53.	a	b	c	d		73.	a	b	c	d
54.	a	b	c	d		74.	a	b	c	d
55.	a	b	c	d		75.	a	b	c	d
56.	a	b	c	d		76.	a	b	c	d
57.	a	b	c	d		77.	a	b	c	d
58.	a	b	c	d		78.	a	b	c	d
59.	a	b	c	d		79.	a	b	c	d
60.	a	b	c	d		80.	a	b	c	d

81.	a	b	c	d
82.	a	b	c	d
83.	a	b	c	d
84.	a	b	c	d
85.	a	b	c	d
86.	a	b	c	d
87.	a	b	c	d
88.	a	b	c	d
89.	a	b	c	d
90.	a	b	c	d
91.	a	b	c	d
92.	a	b	c	d
93.	a	b	c	d
94.	a	b	c	d
95.	a	b	c	d
96.	a	b	c	d
97.	a	b	c	d
98.	a	b	c	d
99.	a	b	c	d
100.	a	b	c	d

101.	a	b	c	d
102.	a	b	c	d
103.	a	b	c	d
104.	a	b	c	d
105.	a	b	c	d
106.	a	b	c	d
107.	a	b	c	d
108.	a	b	c	d
109.	a	b	c	d
110.	a	b	c	d
111.	a	b	c	d
112.	a	b	c	d
113.	a	b	c	d
114.	a	b	c	d
115.	a	b	c	d
116.	a	b	c	d
117.	a	b	c	d
118.	a	b	c	d
119.	a	b	c	d
120.	a	b	c	d

	a	b	c	d		a	b	c	d
121.	a	b	c	d	141.	a	b	c	d
122.	a	b	c	d	142.	a	b	c	d
123.	a	b	c	d	143.	a	b	c	d
124.	a	b	c	d	144.	a	b	c	d
125.	a	b	c	d	145.	a	b	c	d
126.	a	b	c	d	146.	a	b	c	d
127.	a	b	c	d	147.	a	b	c	d
128.	a	b	c	d	148.	a	b	c	d
129.	a	b	c	d	149.	a	b	c	d
130.	a	b	c	d	150.	a	b	c	d
131.	a	b	c	d	151.	a	b	c	d
132.	a	b	c	d	152.	a	b	c	d
133.	a	b	c	d	153.	a	b	c	d
134.	a	b	c	d	154.	a	b	c	d
135.	a	b	c	d	155.	a	b	c	d
136.	a	b	c	d	156.	a	b	c	d
137.	a	b	c	d	157.	a	b	c	d
138.	a	b	c	d	158.	a	b	c	d
139.	a	b	c	d	159.	a	b	c	d
140.	a	b	c	d	160.	a	b	c	d

161.	a	b	c	d	181.	a	b	c	d
162.	a	b	c	d	182.	a	b	c	d
163.	a	b	c	d	183.	a	b	c	d
164.	a	b	c	d	184.	a	b	c	d
165.	a	b	c	d	185.	a	b	c	d
166.	a	b	c	d	186.	a	b	c	d
167.	a	b	c	d	187.	a	b	c	d
168.	a	b	c	d	188.	a	b	c	d
169.	a	b	c	d	189.	a	b	c	d
170.	a	b	c	d	190.	a	b	c	d
171.	a	b	c	d	191.	a	b	c	d
172.	a	b	c	d	192.	a	b	c	d
173.	a	b	c	d	193.	a	b	c	d
174.	a	b	c	d	194.	a	b	c	d
175.	a	b	c	d	195.	a	b	c	d
176.	a	b	c	d	196.	a	b	c	d
177.	a	b	c	d	197.	a	b	c	d
178.	a	b	c	d	198.	a	b	c	d
179.	a	b	c	d	199.	a	b	c	d
180.	a	b	c	d	200.	a	b	c	d

Appendix: Study Matrix

Overview

Periodically, the Project Management Institute (PMI®) publishes a Role Delineation Study (RDS) and uses to define the responsibilities of the recipients of the Project Management Professional (PMP®) credential, In June 2015, PMI issued a new RDS, for the PMP®, which then was published as the PMI® PMP® *Examination Content Outline* (ECO). You can download it from PMI's web site und the Certification section and then by the PMP©. It serves as the foundation for the PMP® exam and for our 200-question practice test in this book and in our on-line test. This RDS is in effect for this PMP© exam beginning in 2018.

The ECO identified five broad performance domains and determined how the 175 questions on the PMP® exam would be distributed according to these domains.* The distribution is as follows:

I	Initiating	13%
II	Planning	24%
III	Executing	31%
IV	Monitoring and Controlling	25%
V	Closing	7%

The matrix beginning on page identifies each practice test question according to its performance domain and its knowledge area in the *PMBOK® Guide*.

The matrix is designed to help you—

- Assess your strengths and weaknesses in each of the performance domains
- Identify those areas in which you need additional study before you take the PMP® exam

---

* PMI® distributes 25 pretest questions across the five domains in any way that it deems appropriate for the purpose of "testing" the questions.

Here is an easy way to use the matrix:

Step 1   Circle all the questions you missed on the practice test in Column 1.

Step 2   For each circled question, note the corresponding process in Column 2.

Step 3   To determine whether any patterns emerge indicating weak areas, tally the information you obtained from the matrix.

Step 4   To ensure that you have a good understanding of the major management processes that define a particular knowledge area, including the inputs, tools and techniques, and outputs, refer to the appropriate knowledge area in the *PMBOK® Guide*.

The last column in the matrix is provided for your notes.

Study Matrix

Practice Test Question Number	Performance Domain Process	Knowledge Area	Study Notes
1	Planning	Scope	
2	Executing	Stakeholders	
3	Executing	Procurement	
4	Planning	Scope	
5	Planning	Integration	
6	Monitoring and Controlling	Procurement	
7	Executing	Integration	
8	Planning	Integration	
9	Planning	Scope	
10	Executing	Quality	
11	Executing	Quality	
12	Monitoring and Controlling	Procurement	
13	Monitoring and Controlling	Integration	
14	Planning	Procurement	
15	Monitoring and Controlling	Integration	
16	Initiating	Integration	
17	Executing	Quality	
18	Monitoring and Controlling	Integration	
19	Monitoring and Controlling	Integration	
20	Monitoring and Controlling	Integration	
21	Planning	Scope	
22	Executing	Integration	
23	Monitoring and Controlling	Integration	
24	Executing	Quality	
25	Monitoring and Controlling	Scope	
26	Closing	Integration	
27	Executing	Quality	
28	Planning	Risk	
29	Planning	Scope	
30	Executing	Quality	

Practice Test Question Number	Performance Domain Process	Knowledge Area	Study Notes
31	Monitoring and Controlling	Integration	
32	Initiating	Integration	
33	Executing	Resources	
34	Monitoring and Controlling	Schedule	
35	Monitoring and Controlling	Cost	
36	Monitoring and Controlling	Cost	
37	Planning	Cost	
38	Closing	Integration	
39	Planning	Cost	
40	Monitoring and Controlling	Cost	
41	Executing	Resources	
42	Monitoring and Controlling	Cost	
43	Monitoring and Controlling	Cost	
44	Initiating	Integration	
45	Planning	Integration	
46	Executing	Integration	
47	Monitoring and Controlling	Cost	
48	Planning	Scope	
49	Executing	Communications	
50	Monitoring and Controlling	Cost	
51	Executing	Resources	
52	Planning	Integration	
53	Monitoring and Controlling	Schedule	
54	Planning	Risk	
55	Executing	Resources	
56	Monitoring and Controlling	Cost	
57	Monitoring and Controlling	Cost	
58	Initiating	Integration	
59	Closing	Integration	
60	Planning	Risk	
61	Executing	Resources	
62	Planning	Schedule	

Practice Test Question Number	Performance Domain Process	Knowledge Area	Study Notes
63	Monitoring and Controlling	Communications	
64	Planning	Integration	
65	Executing	Stakeholders	
66	Planning	Schedule	
67	Monitoring and Controlling	Stakeholders	
68	Monitoring and Controlling	Integration	
69	Planning	Stakeholders	
70	Planning	Integration	
71	Planning	Schedule	
72	Planning	Schedule	
73	Monitoring and Controlling	Schedule	
74	Executing	Integration	
75	Executing	Resources	
76	Closing	Integration	
77	Monitoring and Controlling	Procurement	
78	Initiating	Integration	
79	Monitoring and Controlling	Quality	
80	Executing	Resources	
81	Monitoring and Controlling	Quality	
82	Executing	Resources	
83	Monitoring and Controlling	Quality	
84	Monitoring and Controlling	Quality	
85	Executing	Resources	
86	Monitoring and Controlling	Schedule	
87	Executing	Integration	
88	Initiating	Integration	
89	Executing	Communications	
90	Monitoring and Controlling	Resources	
91	Monitoring and Controlling	Resources	
92	Planning	Cost	
93	Planning	Cost	
94	Executing	Resources	

Practice Test Question Number	*Performance Domain Process*	*Knowledge Area*	*Study Notes*
95	Closing	Integration	
96	Monitoring and Controlling	Scope	
97	Planning	Schedule	
98	Executing	Stakeholders	
99	Monitoring and Controlling	Stakeholders	
100	Planning	Communications	
101	Planning	Schedule	
102	Planning	Schedule	
103	Planning	Communications	
104	Executing	Resources	
105	Initiating	Integration	
106	Monitoring and Controlling	Risk	
107	Initiating	Integration	
108	Closing	Integration	
109	Planning	Stakeholders	
110	Executing	Stakeholders	
111	Closing	Integration	
112	Closing	Integration	
113	Closing	Integration	
114	Initiating	Integration	
115	Executing	Resources	
116	Planning	Resources	
117	Executing	Resources	
118	Monitoring and Controlling	Resources	
119	Initiating	Integration	
120	Monitoring and Controlling	Risk	
121	Initiating	Integration	
122	Monitoring and Controlling	Risk	
123	Monitoring and Controlling	Risk	
124	Monitoring and Controlling	Procurement	
125	Monitoring and Controlling	Procurement	
126	Planning	Quality	

Practice Test Question Number	Performance Domain Process	Knowledge Area	Study Notes
127	Initiating	Integration	
128	Executing	Resources	
129	Planning	Stakeholders	
130	Initiating	Stakeholders	
131	Executing	Integration	
132	Executing	Resources	
133	Planning	Schedule	
134	Executing	Resources	
135	Initiating	Stakeholders	
136	Closing	Integration	
137	Monitoring and Controlling	Procurement	
138	Closing	Integration	
139	Executing	Resources	
140	Monitoring and Controlling	Communications	
141	Monitoring and Controlling	Scope	
142	Executing	Resources	
143	Executing	Resources	
144	Initiating	Stakeholders	
145	Executing	Communications	
146	Executing	Risk	
147	Initiating	Stakeholders	
148	Executing	Integration	
149	Planning	Cost	
150	Initiating	Integration	
151	Closing	Integration	
152	Monitoring and Controlling	Scope	
153	Initiating	Integration	
154	Initiating	Stakeholders	
155	Initiating	Stakeholders	
156	Executing	Integration	
157	Planning	Resources	
158	Initiating	Stakeholders	

Practice Test Question Number	*Performance Domain Process*	*Knowledge Area*	*Study Notes*
159	Initiating	Stakeholders	
160	Planning	Communications	
161	Planning	Quality	
162	Planning	Quality	
163	Executing	Procurement	
164	Executing	Risk	
165	Executing	Procurement	
166	Planning	Procurement	
167	Planning	Resources	
168	Planning	Resources	
169	Planning	Resources	
170	Executing	Integration	
171	Closing	Integration	
172	Executing	Communications	
173	Executing	Integration	
174	Planning	Integration	
175	Initiating	Stakeholders	
176	Executing	Communications	
177	Executing	Resources	
178	Initiating	Stakeholders	
179	Planning	Risk	
180	Executing	Integration	
181	Executing	Procurement	
182	Executing	Stakeholders	
183	Planning	Risk	
184	Closing	Integration	
185	Closing	Integration	
186	Planning	Risk	
187	Planning	Risk	
188	Monitoring and Controlling	Communications	
189	Executing	Procurement	
190	Monitoring and Controlling	Risk	

Practice Test Question Number	Performance Domain Process	Knowledge Area	Study Notes
191	Planning	Procurement	
192	Monitoring and Controlling	Scope	
193	Executing	Integration	
194	Executing	Integration	
195	Planning	Procurement	
196	Initiating	Stakeholders	
197	Executing	Integration	
198	Executing	Procurement	
199	Executing	Integration	
200	Initiating	Stakeholders	

Answer Key

1. c. It is not a problem at this time. The previous project manager was using the rolling wave planning technique, so you are able to continue defining the activities.

    Rolling wave planning provides progressive detailing of the work to be accomplished throughout the life of the project. Decomposition is a tool and technique in Create WBS and recognizes rolling wave planning may be needed for deliverables or subcomponents that will not be accomplished until much later in the project. In this case, the team waits until more information about the deliverable or subcomponent is available. This approach is rolling wave planning. [Planning]

    PMI®, *PMBOK® Guide*, 2017, 160
    PMI® *PMP Examination Content Outline,* 2015, Planning, 6, Task 2

2. a. You want to increase their support and minimize resistance

    During the Manage Stakeholder Engagement process, its benefit is that it enables the project manager to increase support and minimize resistance from stakeholders. This process involves communicating and working with stakeholders to meet their needs and expectations. [Executing]

    PMI®, *PMBOK® Guide*, 2017, 523
    PMI® *PMP Examination Content Outline,* 2015, Executing, 8, Task 7

3. a. Review invoicing and payment information

    Invoicing and payment information as well as financial policies and procedures are one of the organizational process assets to review as an input to Conduct Procurements. The selected seller(s) will want to know how to submit invoices and also know when they will be paid, often dictated by the terms and conditions in the type of agreement. [Executing]

    PMI®, *PMBOK® Guide*, 2017, 486
    PMI® *PMP Examination Content Outline*, 2015, Executing, 8, Task 1

4. c. Determining problems and identifying business needs

    A greater focus on requirements is an emerging trend in project management. Increasingly in the global world in which project work is done, more organizations are using business analysts to help define, manage, and control requirements activities. A business analyst is assigned to this team in this question. For success, the business analyst and the project manager need to work collaboratively and understand each other's role. Through this collaboration, the first step is to determine the problems and understand the business needs. It is followed by identifying and recommending solutions. Then, the requirements are elicited, documented, and managed, and finally the goal is to implement them successfully. [Planning]

    PMI®, *PMBOK® Guide*, 2017, 132
    PMI® *PMP Examination Content Outline*, 2015, Planning, 6, Task 2

5. c. Part of the project management plan

    Most of the components in the project management plan are produced in other processes than in Project Integration Management. These three are ones that are prepared during this part of Project Integration Management. The project life cycle describes the phases in the project from initiating to closing; the development approach describes whether it is one that is predictive, iterative, agile, or a hybrid; and the management reviews are the times in the project in which the project manager and other stakeholders will review progress to see if performance is as expected or if preventive or corrective action is needed. [Planning]

    PMI®, *PMBOK® Guide*, 2017, 88
    PMI® *PMP Examination Content Outline*, 2015, Planning, 7, Task 11

6. a. Control Procurements

   The purpose of Control Procurements is to ensure that the contractual requirements are met by the seller. This objective is accomplished by managing procurement relationships, monitoring contract performance, making changes and corrections to contracts if appropriate, and closing contracts. In this process, change requests are an output and are processed for review and disposition through the Perform Integrated Change Control process. [Monitoring and Controlling]

   PMI®, *PMBOK® Guide*, 2017, 499
   PMI® *PMP Examination Content Outline*, 2015, Monitoring and Controlling, 9, Task 7

7. b. The lowest level of detail to derive information

   Work performance data are an output of the Direct and Manage Project Work process. They are the raw observations and measurements identified during activities being performed to do the project work. Therefore, they are often viewed as the lowest level of detail from which to derive information by other processes. The data are gathered and then passed to the controlling processes of each process area for further analysis. [Executing]

   PMI®, *PMBOK® Guide*, 2017, 95
   PMI® *PMP Examination Content Outline*, 2015, Executing, 7, Task 2

8. a. Use it to involve your team in the planning process

   Meetings are a tool and technique in the Develop Project Management Process when the project management plan is prepared. A kick-off meeting tends to be used when the plan is complete, and the executing process begins. However, in this situation, it is a small project. The kick-off meeting thus is recommended shortly after initiating is complete since there is usually only one team responsible for planning and executing. [Planning]

   PMI®, *PMBOK® Guide*, 2017, 66
   PMI® *PMP Examination Content Outline*, 2015, Planning, 7, Task 11

9. c. Job shadowing

   Observation/conversation are an interpersonal and team skill tool and technique in the Collect Requirements process. They provide a way to view individuals in their environment and to see how they perform their jobs or tasks and carry out processes. It is helpful in situations in which people may have difficulty or are reluctant to state their requirements. Another term for this approach is job shadowing and usually is done by an observer viewing the person performing his or her job. It can also be done by a 'participant observer' who is performing a process or procedure to experience how it is done to uncover hidden requirements. [Planning]

   PMI®, *PMBOK® Guide*, 2017, 145
   PMI® *PMP Examination Content Outline*, 2015, Planning, 6, Task 2

10. a. Plan-do-check-act

    The plan-do-check act cycle is the basis for quality improvement and efficiency. It was defined by Shewhart and modified by Deming. Quality Management focuses on quality improvement, which also includes total quality management, Six Sigma, and Lean Six Sigma. These techniques improve project management quality and the quality of the product, service, or result. Quality improvement methods are a tool and technique in Manage Quality. [Executing]

    PMI®, *PMBOK® Guide*, 2017, 275, 296
    PMI® *PMP Examination Content Outline*, 2015, Executing, 8, Task 3

11. c. Use an affinity diagram

    In Manage Quality, an affinity diagram is a data representation tool and technique. This diagram is used to generate ideas that can be linked to form organized patterns of thought about a problem. Using them in project management, one can enhance the creation of the WBS by using it to give structure to the decomposition of scope. [Executing]

    PMI®, *PMBOK® Guide*, 2017, 293
    PMI® *PMP Examination Content Outline*, 2015, Executing, 8, Task 3

12. d. Managing the interfaces among the sellers

> The Control Procurements process involves managing procurement relationships, monitoring contract performance, and making changes and corrections to contracts as needed. On a large project with multiple sellers as in this question, a major aspect is managing the interfaces among the sellers. Many organizations have contract administration as an administrative function separate from the project organization. While a procurement administrator may be on the core team, typically this person reports to someone other than the project manager in a different department. [Monitoring and Controlling]

> PMI®, *PMBOK® Guide*, 2017, 494
> PMI® *PMP Examination Content Outline*, 2015, Monitoring and Controlling, 9, Task 7

13. c. Perform verification and audit on each configuration item

> Configuration management is an integral part of the Perform Integrated Change Control process. The first step is to identify the specific configuration item. The next step is to record and report the status of each configuration item. The answer is the last step. The verification and audits ensure the project's configuration item is correct. It also ensures any corresponding changes are registered, assessed, approved, tracked and implemented correctly. The purpose is to ensure the functional requirements defined in the formal configuration documents are met. [Monitoring and Controlling]

> PMI®, *PMBOK® Guide*, 2017, 118
> PMI® *PMP Examination Content Outline*, 2015, Monitoring and Controlling, 8, Task 2

14. b. Procurement management plan

> The procurement management plan contains the activities to be done in the procurement process regardless of whether it is an international, national, or local bid. If it is financed externally, the funding sources and availability should be aligned with the procurement management plan and project schedule. It may be formal or informal, detailed or broadly framed based on the project needs. [Planning]

> PMI®, *PMBOK® Guide*, 2017, 475
> PMI® *PMP Examination Content Outline*, 2015, Planning, 6, Task 7

15. c. Variance analysis

   Variance analysis reviews the differences or variances between planned and actual performance it also can be conducted in each of the Knowledge Areas, but in Monitor and Control Project Work, it reviews variances from an integrated perspective considering cost, time, and resource variance in relationship to each other. The objective is to have an overview view of variance on the project such that if required preventive or corrective action can be taken. [Monitoring and Controlling]

   PMI®, *PMBOK® Guide*, 2017, 111
   PMI® *PMP Examination Content Outline*, 2015, Monitoring and Controlling, 9, Task 1

16. d. Ecological impact

   The business case is created in this question primarily as a result of an ecological need. Answer b refers to a customer request such as an electrical utility authorizing a project to build a new substation to serve a new industrial park. Answer c relates to a social need such as a non-governmental organization in a developing country to provide portable water systems plus other items to communities suffering from high rates of cholera. Other reasons are a market demand, an organizational need, a technological advance, competitive forces, material issues, stakeholder demands, political changes, economic changes, business process improvements, or a legal requirement. Sustainability does not apply. The business case is an input to the Develop Project Charter process. [Initiating]

   PMI®, *PMBOK® Guide*, 2017, 9, 78
   PMI® *PMP Examination Content Outline*, 2015, Initiating, 4, Task 1

17. d. A quality audit

    A quality audit is a tool and technique in the Manage Quality process. It is primarily used to determine whether the project team is complying with organizational and project policies, processes, and procedures. It identifies good and best practices being implemented; areas of nonconformity, gaps, and shortcomings; good practices introduced or implemented in similar projects or in the industry; ways to offer assistance positively to improve process implementation to help the team increase productivity; and highlights of the audit are placed in the lessons-learned repository. [Executing]

    PMI®, *PMBOK® Guide*, 2017, 247
    PMI® *PMP Examination Content Outline,* 2015, Executing, 7, Task 3

18. b. Defect repairs

    Change requests are an input to the Perform Integrated Change Control process. They also may include corrective action and preventive action and updates to formally controlled documents or deliverables if they reflect additional ideas or content. Sometimes performance against baselines may be affected, but the project manager is responsible for making these decisions. If the change requests do affect baselines, additional information is required. [Monitoring and Controlling]

    PMI®, *PMBOK® Guide*, 2017, 117
    PMI® *PMP Examination Content Outline,* 2015, Monitoring and Controlling, 9, Task 2

19. d. Identified in the project management plan

    Usually, the project manager or the project sponsor can approve or reject a documented change request. At times, a Change Control Board (CCB) is used, which later may require customer or sponsor approval unless they are on the CCB. Regardless, the responsible person is identified in the project management plan or by organizational procedures. [Monitoring and Controlling]

    PMI®, *PMBOK® Guide*, 2017, 96
    PMI® *PMP Examination Content Outline,* 2017, Monitoring and Controlling, 9, Task 1

20. d. Change control meetings

Meetings, referred to as change control meetings, are a tool and technique in Integrated Change Control. Often, a project will set up a CCB, which has the responsibility for meeting and reviewing the change requests, and approving, rejecting, or deferring them. Assessing the impact of the change is part of the meeting, and alternatives may be discussed and proposed. Finally, the decision of the CCB is communicated to the required user or group. The CCB also may review configuration management activities. Decisions of the board are documented and agreed upon by appropriate stakeholders and are documented in the change management plan. The CCB decisions are documented to stakeholders for information and follow-up actions. [Monitoring and Controlling]

PMI®, *PMBOK® Guide*, 2017, 120
PMI® *PMP Examination Content Outline*, 2015, Monitoring and Controlling, 9, Task 2

21. b. Adds business value as it links to business and project objectives

The requirements traceability matrix is a grid that links requirements to their origin and traces them throughout the life cycle. It is an output in Collect Requirements. This approach helps to ensure that each requirement adds value as it links to the business and project objectives. It also tracks requirements during the life cycle to help ensure that the requirements listed in the requirements document are delivered at the end of the project. It provides a structure to manage changes to product scope. [Planning]

PMI®, *PMBOK® Guide*, 2017, 148
PMI® *PMP Examination Content Outline*, 2015, Planning, 6, Task 2

22. b. Enterprise environmental factors

They are an input to the Direct and Manage Project Work process, and as an example, the stakeholder risk tolerances are allowable percent of cost overruns. Other examples of enterprise environmental factors are organizational, cultural, management practices and sustainability; and infrastructure such as existing facilities and capital equipment. [Executing]

PMI®, *PMBOK® Guide*, 2017, 93
PMI® *PMP Examination Content Outline*, 2015, Executing, 7, Task 7

23. c. Requirements traceability matrix

    Three project documents to consider since they are inputs to Perform Integrated Change Control are the basis of estimates [another possible answer to this question], the risk report, and the requirements traceability matrix. It is useful because it helps assess the impact of the change on the project's scope. [Monitoring and Controlling]

    PMI®, *PMBOK® Guide*, 2017, 116
    PMI® *PMP Examination Content Outline*, 2015, Monitoring and Controlling, 9, Task 2

24. b. Concentrating on small batch systems

    Agile methods involve frequent quality and review steps that are built into the project. They focus on small batch systems in order to uncover inconsistencies and quality issues as early in the project as possible when the overall life cycle costs are low. A number of retrospectives also are used to evaluate the effectiveness of the quality processes, so improvement can be made. [Executing]

    PMI®, *PMBOK® Guide*, 2017, 276
    PMI® *PMP Examination Content Outline*, 2015, Executing, 8, Task 3

25. c. Control Quality is concerned with meeting the quality requirements for the deliverables

    Validate Scope focuses on accepting project deliverables. Control Quality is one way to ensure the correctness of the deliverables and meeting the quality specifications for the deliverables, which is why Control Quality typically is done before Validate Scope. Further, the verified deliverables obtained from Control Quality have been reviewed with the customer or sponsor to ensure they are completely satisfied and have received formal acceptance from the customer or sponsor. In the Validate Scope process, verified deliverables are a tool and technique used to ensure project deliverables are completed and checked for correctness through Control Quality. [Monitoring and Controlling]

    PMI®, *PMBOK® Guide*, 2017, 164–165
    PMI® *PMP Examination Content Outline*, 2015, Monitoring and Controlling, 8, Task 3

26. c. Making sure all the activities necessary to satisfy exit criteria for a phase or the entire project are followed

    Administrative activities are necessary to close the project or a phase. The answer is the first activity to do to satisfy completion of the exit criteria for the phase or the project. The second administrative activity is to ensure contractual agreements are completed; followed by transferring the project's products, services, or results to the next phase or productions or operations; collecting suggestion for improving organizational policies and procedures; and last but not least measuring stakeholder satisfaction. [Closing]

    PMI®, *PMBOK® Guide*, 2017, 123
    PMI® *PMP Examination Content Outline*, 2015, Closing, 10, Task 7

27. c. Finding solutions for these issues and other challenges

    Problem solving is a tool and technique in Manage Quality. It involves finding solutions for issues or challenges. It includes gathering additional information, using critical thinking, being creative, and using quantitative and logical approaches. Often, problems can arise from Control Quality or through audits. They may be associated with a process or a deliverable. Problem solving with a structured approach tends to result in a long-lasting solution. [Executing]

    PMI®, *PMBOK® Guide*, 2017, 295
    PMI® *PMP Examination Content Outline*, 2015, Executing, 8, Task 3

28. b. You are holding bi-weekly reviews to identify emergent risks.

    Project resilience is an emerging trend in risk management with the increase in unknown unknowns. These risks are recognized after they occur, which means the best way to handle them is through project resilience. One way to do so especially in the Monitor Risks process is to have frequent reviews of early warning signs such that any emergent risks can be identified and then managed as soon as possible. [Monitoring and Controlling]

    PMI®, *PMBOK® Guide*, 2017, 399
    PMI® *PMP Examination Content Outline*, 2015, Monitoring and Controlling, 9, Task 4

29. a. Need to spend less time defining requirements early on your project

In an agile environment, especially if the requirements are evolving, the scope may not be understood at the beginning of the project. Using agile methods then less time is spent trying to define and agree on scope early in the project. More time instead is spent in establishing the process for ongoing discovery and refinement. This approach tends to mean there is a gap between the real business requirements and the business requirements that were stated when the project was approved. With agile, it focuses on building and reviewing prototypes and release versions to refine requirements. Scope then is defined and redefined throughout the project, and requirements constitute the backlog. [Planning]

PMI®, *PMBOK® Guide*, 2017, 133
PMI® *PMP Examination Content Outline*, 2015, Planning, 6, Task 2

30. b. Work shadowing

There are a number of tools and techniques in Manage Project Knowledge. Some key words in this question are diversity, innovation, and complexity. Therefore, since the project represents a new way of working, work shadowing is an excellent tool and technique to use. In it you or a member of your team can observe how a key stakeholder does his or her work under the current policies and procedures to learn what they will need to do to change to the new methods and the person's level of resistance involved. The other answers are from inputs to this process or other tools and techniques in it. [Executing]

PMI®, *PMBOK® Guide*, 2017, 103
PMI® *PMP Examination Content Outline*, 2015, Executing, 8, Task 6

31. d. Continuously monitor the project

The Monitor and Control Project Work process is performed throughout the project. Monitoring includes collecting, measuring, and disseminating performance information and assessing measurements and trends to effect process improvement. Continuous monitoring is important because it provides insight into the project's health, highlighting areas that may require special attention. [Monitoring and Controlling]

PMI®, *PMBOK® Guide*, 2017, 107
PMI® *PMP Examination Content Outline*, 2015, Monitoring and Controlling, 9, Task 1

32. d. Determine whether the project should continue to the next phase

    The review at the end of a project phase is called a phase gate. It also may be called a stage gate, phase review, kill point, or phase entrance or phase exit. In this review, the project's performance and progress are compared to the project and business documents such as the business case, project charter, project management plan, and the benefits management plan. The purpose of this review is to determine whether the project should continue to the next phase, continue with modification, end the project, stay in the phase, or repeat the entire phase or parts of it. [Initiating]

    PMI®, *PMBOK® Guide*, 2017, 21
    PMI® *PMP Examination Content Outline*, 2015, Initiating, 4, Task 5

33. b. Empower the team

    In the performing stage of team development according to the Tuckman model, the team functions as a well-organized group. They work interdependently and work through issues smoothly and effectively. The project manager then can empower the team to participate in decision making and take ownership of the decisions to improve team productively and achieve more effective and efficient results. [Executing]

    PMI®, *PMBOK® Guide*, 2017, 338
    PMI® *PMP Examination Content Outline*, 2015, Executing, 8, Task 2

34. b. At +$300, the situation is favorable, as physical progress is being accomplished ahead of your plan.

    Schedule variance is calculated as EV − PV, or $1,500 − $1,200 = +$300. Because the SV is positive, physical progress is being accomplished at a faster rate than planned. It is a useful metric as it can indicate when a project is failing behind or is ahead of its baseline schedule and should be used along with critical path analysis. [Monitoring and Controlling]

    PMI®, *PMBOK® Guide*, 2017, 226
    PMI® *PMP Examination Content Outline*, 2015, Monitoring and Controlling, 8, Task 1

35. b. Ask open-ended questions and use active listening

The Monitor Stakeholder Engagement process is ongoing throughout the project. This situation shows a leading stakeholder, the Chief Financial Officer, has lost interest. The process suggests a number of interpersonal and team skills to use, one of which is active listening. At the meeting with this key stakeholder, the project manager should formulate some open-ended questions to ask. Then, he or she uses active listening to reduce any misunderstandings and any miscommunication. Since the goal is to continue to ensure this stakeholder is in the leading category, as a next step the project manager should request another meeting and explain strategies to modify the stakeholder engagement strategy to accommodate the Chief Financial Officer's concerns. [Monitoring and Controlling]

PMI®, *PMBOK® Guide*, 2017, 534
PMI® *PMP Examination Content Outline*, 2015, Monitoring and Controlling, 9, Task 1

36. d. Recognize that your original estimates were fundamentally flawed, and your project is in an atypical situation

CPI = EV/AC. It is considered the most critical earned value management metric since it measures the cost efficiency for the completed work. The CPI is useful for determining project status and provides a basis to estimate project cost and schedule outcomes. A CPI of 0.44 means that for every dollar spent, you are only receiving 44 cents of progress. Therefore, something is not correct with how you planned your project, or your original estimates were fundamentally flawed, and your project is in an atypical situation. You might want to reconsider a formal "replan" and/or take a new baseline of your project. [Monitoring and Controlling]

PMI®, *PMBOK® Guide*, 2017, 263
PMI® *PMP Examination Content Outline*, 2015, Monitoring and Controlling, 9, Task 1

37. c. A high-level cost forecast

Projects with a high degree of uncertainty or if the scope is not fully defined may not benefit from detailed cost calculations because there are frequent changes. In an agile environment, lightweight estimates can be used to generate a high-level cost of project labor costs, which can be adjusted when changes are made. A detailed estimate is then used for a short-term planning horizon. [Planning]

PMI®, *PMBOK® Guide*, 2017, 235
PMI® *PMP Examination Content Outline*, 2015, Planning, 6, Task 3

38. c. Expert judgment

    Expert judgment is used as a tool and technique in Close Project or Phase to consider individuals or groups with specialized knowledge or training in management control, audit, legal and procurement, and legislation and regulation. [Closing]

    PMI®, *PMBOK® Guide*, 2017, 126
    PMI® *PMP Examination Content Outline*, 2015, Closing, 10, Task 1

39. b. Have imposed date constraints

    In the Develop Budget process, funding limit reconciliation is a tool and technique. In it the expenditure of funds is reconciled with the commitment of funds for the project. If there is a variance between the funding limits and the planned expenditures, it may mean the rescheduling of some work to level out the expenditures. It is accomplished by placing imposed date constraints on work in the project's schedule. [Planning]

    PMI®, *PMBOK® Guide*, 2017, 253
    PMI® *PMP Examination Content Outline*, 2015, Planning, 6, Task 3

40. a. Informing your stakeholders of approved change requests and their costs

    A number of activities are involved in the Control Costs process. The key to its effectiveness is in managing the approved cost baseline and any changes to it. Since change is inevitable on projects, it is a best practice each time there is an approved change to notify stakeholders of the costs to implement it, which tracks to this question as management is interested in a focus on controlling costs. Often stakeholders wish to see a cost forecast using the EAC. [Monitoring and Controlling]

    PMI®, *PMBOK® Guide*, 2017, 209
    PMI® *PMP Examination Content Outline*, 2015, Monitoring and Controlling, 9, Task 1

41. a. Commence following guidelines in the resource management plan

    Training is a tool and technique for the Develop Project Team process. The resource management plan has a section on training in terms of training for team members. Training includes activities designed to enhance team member competencies, as in this example in contract administration. A variety of types of training methods can be used. Scheduled training takes place according to the resource management plan. Unplanned training also can be used. [Executing]

    PMI®, *PMBOK® Guide*, 2017, 342
    PMI® *PMP Examination Content Outline*, 2015, Executing, 8, Task 2

42. d. The range of acceptable variances will tend to decrease over time

    Variance analysis, along with trend analysis, is examples of performance reviews, a tool and technique in Control Costs. The most frequently analyzed measurements are cost and schedule variances. Cost performance measurements are used to assess the magnitude of variation to the original cost baseline. It is necessary to determine the cause and degree of variance relative to the baseline, but over time, the percentage range of acceptable variation will decrease as stakeholders see more work is being accomplished. They also will be less likely to terminate the project recognizing its sunk costs and the work completed to date. [Monitoring and Controlling]

    PMI®, *PMBOK® Guide*, 2017, 262
    PMI® *PMP Examination Content Outline*, 2015, Monitoring and Controlling, 9, Task 1

43. c. Estimate at completion

    EAC is the expected total cost to complete all work expressed as the sum of the actual costs to date and the estimate to complete [ETC] or the expected cost to finish all remaining work. It can be calculated several different ways. To use it effectively the project team must predict what it will take for the ETC based on experience to date. It may differ from the budget at completion. It is a data analysis tool and technique in this process. [Monitoring and Controlling]

    PMI®, *PMBOK® Guide*, 2017, 264–265
    PMI® *PMP Examination Content Outline*, 2015, Monitoring and Controlling, 9, Task 1

44. d. Links the project to the organization's strategic objectives

    The project charter not only authorizes a project, it shows how the project is linked to the strategic objectives of the organization, creates an ongoing record of the project, and shows the organization is committed to the project. [Initiating]

    PMI®, *PMBOK® Guide*, 2017, 75
    PMI® *PMP Examination Content Outline*, 2015, Initiating, 5, Task 1

45. a. Conflict management

Interpersonal and team skills are a tool and technique in the Develop Project Management Plan process. Other examples are facilitation and meeting management. In conflict management the objective is to bring diverse stakeholders into alignment on all parts of the project management plan. [Planning]

PMI®, *PMBOK® Guide*, 2017, 86
PMI® *PMP Examination Content Outline*, 2015, Planning, 7, Task 11

46. d. Performance measurement data base

The performance measurement data base is an organizational process asset that is used to collect and make available measurement data on processes and products. The other answers are examples of enterprise environmental factors or expert judgment, other inputs to Direct and Manage Project Work. [Executing]

PMI®, *PMBOK® Guide*, 2017, 94
PMI® *PMP Examination Content Outline*, 2015, Executing, 7, Task 2

47. c. EAC = AC + Bottom-up ETC

This formula assumes that all of the remaining work is independent of the burn rate incurred thus far. AC is $2,900 + [$500 + $1,000]. The $500 is from Activity B, and the $1,000 is from Activity C. This bottom-up approach builds on the actual costs and experience incurred for the work completed and requires a new estimate to complete the remaining work. [Monitoring and Controlling]

PMI®, *PMBOK® Guide*, 2017, 264–265
PMI® *PMP Examination Content Outline*, 2015, Monitoring and Controlling, 9, Task 1

48. c. Use prioritization/ranking

Decision making is a tool and technique in Plan Stakeholder Engagement. Prioritization/ranking is used as an approach to ensure stakeholder requirements are prioritized and ranked. The purpose is to make sure the stakeholders with the most interest and highest influence are prioritized at the top of the list. [Planning]

PMI®, *PMBOK® Guide*, 521
PMI® *PMP Examination Content Outline*, 2015, Planning, 7, Task 13

49. c. Manage Communications process

Lessons learned documentation is an output of the Manage Communications process as the lessons-learned register. It includes information on communications challenges encountered and how they could have been avoided and approaches that worked well and areas that need improvement in managing communications. [Executing]

PMI®, *PMBOK® Guide*, 2017, 387
PMI® *PMP Examination Content Outline*, 2015, Executing, 8, Task 6

50. a. Influencing the factors that create change to the authorized cost baseline

The Control Costs process is concerned with ensuring that requested changes have been acted upon, managing actual changes if and when they occur, ensuring cost expenditures do not exceed authorized funding, monitoring cost and work performance, preventing unapproved changes from being included in the reported cost or resource use, informing stakeholders of all approved changes and their costs, and bringing expected cost overruns within acceptable limits. [Monitoring and Controlling]

PMI®, *PMBOK® Guide*, 2017, 259
PMI® *PMP Examination Content Outline*, 2015, Monitoring and Controlling, 9, Task 1

51. a. Agreement and trust

In the Acquire Team process, the project manager is identifying, motivating, maintaining, leading and inspiring the team to meet project objectives. Teamwork is a critical factor in project success. High team performance is characterized by agreement and trust among team members. The project manager can promote working relationships in a climate of mutual trust. Feelings of trust and agreement among the team members raises morale, lowers conflict, and increases teamwork. [Executing]

PMI®, *PMBOK® Guide*, 2017, 337–338
PMI® *PMP Examination Content Outline*, 2015, Executing, 8, Task 1

52. b. An enterprise environmental factor

Enterprise environmental factors can influence how the project management plan is developed. While there are several to consider, one is the project management body of knowledge for the vertical market such as construction or a focus area - in this case agile software development. Other examples in this category are environmental, safety, or risk as a focus area. [Planning]

PMI®, *PMBOK® Guide*, 2017, 84
PMI® *PMP Examination Content Outline*, 2015, Planning, 7, Task 11

53. b. Reprioritizing the remaining work plan

    There are a number of areas in Control Schedule if agile is used. While one is the answer to this question, others are: comparing the project's current status by comparing the total amount of work delivered and accepted against the estimated amounts of work for the time cycle; conducting retrospectives or lessons learned to correct processes and possibly make improvements; determining the rate deliverables are produced, validated, and accepted in a given time per iteration; determining if the project schedule has changed; and managing actual changes as they occur. [Monitoring and Controlling]

    PMI®, *PMBOK® Guide*, 2017, 224
    PMI® *PMP Examination Content Outline*, 2015, Monitoring and Controlling, 9, Task 1

54. c. $6.42 million

    Test: $5M + $960K + $460K = $6.42M; Don't Test: $7M. Decision tree analysis is a tool and technique as part of Perform Quantitative Risk Analysis to elect the best of several alternative courses of action. It is evaluated by calculating the expected monetary value. [Planning]

    PMI®, *PMBOK® Guide*, 2017, 435
    PMI® *PMP Examination Content Outline*, 2015, Planning, 5, Task 10

55. b. Encourage independent work

    In the Develop Team process, motivation is one of the interpersonal and team skills to use. Although people are motivated differently, the objective of motivation in this process is to provide a reason for someone to act. Teams are motivated if they can be empowered to participate in decision making and if they are encouraged to work independently. [Executing]

    PMI®, *PMBOK® Guide*, 2017, 341
    PMI® *PMP Examination Content Outline*, 2015, Executing, 8, Task 2

56. a. Work performance information

    The project's work performance information includes information as to how the project work is performing compared to the cost baseline. Variances in the work performed and the cost of the work are evaluated at the work package and the control account levels. It documents and communicates the CV, CPI, EAC, TCPI, and VAC in work performance reports. It is an output of Control Costs. [Monitoring and Controlling]

    PMI®, *PMBOK® Guide*, 2017, 268
    PMI® *PMP Examination Content Outline*, 2015, Monitoring and Controlling, 9, Task 1

57. d. A calculated EAC value or a bottom-up EAC value is documented and communicated to stakeholders

     Cost forecasts are another output of Control Costs, and the EAC is used to show the expected total costs of completing all work expressed as the sum of the actual cost to date and the estimate to complete. The EAC is a calculated value or a bottom-up EAC is used. [Monitoring and Controlling]

     PMI®, *PMBOK® Guide*, 2017, 269
     PMI® *PMP Examination Content Outline,* 2015, Monitoring and Controlling, 9, Task 1

58. a. Benefit management plan

     The other answers are covered in the business case and its development. The benefit management plan is critical as it describes how and when the project's benefits will be delivered and ways to measure them. Benefits are outcomes of actions, behaviors, products, services, or results, which provide value to the organization and the project's intended beneficiaries. Developing this plan is done early in the project's life cycle to define the target benefits to be realized. [Initiating]

     PMI®, *PMBOK® Guide*, 2017, 69
     PMI® *PMP Examination Content Outline,* 2015, Initiating, 4, Task 7

59. a. A form of data analysis

     Data analysis techniques are tools and techniques used in Close Project or Phase. Trend analysis is an example of one approach as it can be used to validate the models used in the organization to improve performance on future projects. [Closing]

     PMI®, *PMBOK® Guide*, 2017, 126
     PMI® *PMP Examination Content Outline,* 2015, Closing, 10, Task 5

60. b. The risk appetite

     The risk appetite is an output from Plan Risk Management. It is also recorded in the risk management plan. It is significant as it states the stakeholders' measurable risk thresholds around each project objective. These individual thresholds then determine the acceptable level of overall risk thresholds around a project objective. They also are used in defining the probability and impact definitions used to evaluate and prioritize individual project risks. [Planning]

     PMI®, *PMBOK® Guide*, 2017, 407
     PMI® *PMP Examination Content Outline,* 2015, Planning, 6, Task 10

61. b. Achieving project objectives

    Team assessments are an output of Develop Team. After some development efforts, such as training, and team building are implemented, the project manager conducts informal or formal team assessments of the team's effectiveness. Expected team development activities are supposed to increase the team's performance, which then leads to the likelihood of meeting the project's objectives. [Executing]

    PMI®, *PMBOK® Guide*, 2017, 343
    PMI® *PMP Examination Content Outline*, 2015, Executing, 8, Task 2

62. b. Monte Carlo analysis

    Simulation is a tool and technique for the Develop Schedule process as it models the combined effects of individual project risks and any other sources of uncertainty to help evaluate the probability of achieving the project's objectives. Monte Carlo analysis is the most commonly used simulation technique. When it is used, risks and other sources of uncertainty are used to calculate possible schedule outcomes for the project. It calculates multiple work package durations with different sets of activity assumptions, constraints, issues, or scenarios using probability durations to determine the probability of achieving a certain project finish date. [Planning]

    PMI®, *PMBOK® Guide*, 2017, 213–214
    PMI® *PMP Examination Content Outline*, 2015, Planning, 6, Task 4

63. d. Establish a knowledge management repository

    Even if the origination has a defined project management methodology to follow, each project is unique. A tailoring consideration is knowledge management. It is useful in the Monitor Communications process to set up a knowledge repository. Setting it up is the easy part. For it to be used, it needs up-to-date content that is easy to locate. Also, in Monitor Communications, the project management information system is a tool and technique. The knowledge repository can be part of it since it consists of standard tools for the project manager to capture, store, and distribute information to stakeholders following guidelines in the communications management plan. [Monitoring and Controlling]

    PMI®, *PMBOK® Guide*, 2017, 365, 382
    PMI® *PMP Examination Content Outline*, 2015, Monitoring and Controlling, 9, Task 1

64. d. Configuration management plan

This plan is one that is prepared during the Develop Project Management Plan process and is an additional component to the plan. It is used to describe how information about the project items, and which items, will be recorded and updated. The purpose is to ensure the product, service, or result of the project is consistent and/or operative. [Planning]

PMI®, *PMBOK® Guide*, 2017, 88
PMI® *PMP Examination Content Outline*, 2015, Planning, 7, Task 11

65. c. Effective and ineffective approaches

In the Manage Stakeholder Engagement process, a lessons-learned register is an output. This register is updated with both effective and ineffective approaches used in managing stakeholder engagements. It purpose also is to use it on this project and future projects. [Executing]

PMI®, *PMBOK® Guide*, 2017, 539
PMI® *PMP Examination Content Outline*, 2015, Executing, 8, Task 7

66. b. Doubling the number of needed resources

The Estimate Activity Duration process provides an estimate of the amount of work to complete an activity and the resources estimated to do so. In this question, you know you need key SMEs to be available at certain times, and resources tend to be scarce when specialization is concerned. In estimating durations, the number of resources needed must be considered. While increasing the number of resources to twice the original number can be done, it may not always reduce the time by half; it may even require extra time because of risk. Additionally, at some point too many resources may even increase duration because of knowledge transfer, the learning curve, additional coordination, and possibly other factors. [Planning]

PMI®, *PMBOK® Guide*, 2017, 197
PMI® *PMP Examination Content Outline*, 2015, Planning, 6, Task 4

67. c. Update the project management plan

    If there are changes to the stakeholder engagement strategy and plans, as an output of the Monitor Stakeholder Engagement process, affected sections of the project management plan are updated because of these changes. Examples are the resource management, communications management, and stakeholder engagement plans. In the resource management plan, team responsibilities for stakeholder engagement may need updates. In the communications management plan, the communications strategy may need updates and in the stakeholder engagement plan, information about the stakeholder community may need updates as different stakeholders are interested in different phases of the project. [Monitoring and Controlling]

    PMI®, *PMBOK® Guide*, 2017, 535
    PMI® *PMP Examination Content Outline*, 2015, Monitoring and Controlling, 9, Task 1

68. c. Updated the project management plan

    In the Monitor and Control Project Work process, updates to the project management plan are one of its outputs. Changes identified in this process may affect the overall project plan. If updates to this plan are needed, they are processed with a change request [another output] through the organization's change control process. [Monitoring and Controlling]

    PMI®, *PMBOK® Guide*, 2017, 97
    PMI® *PMP Examination Content Outline*, 2015, Monitoring and Controlling, 9, Task 2

69. c. Ask some experts

    The key words in the question are new to the company. Expert judgment then is a useful tool and technique in Plan Stakeholder Engagement. While the experts can provide advice on a variety of topics, they are especially helpful in the power and politics structures inside and outside the organization and in the environmental and cultural inside and outside of the organization. [Planning]

    PMI®, *PMBOK® Guide*, 2017, 131
    PMI® *PMP Examination Content Outline*, 2015, Planning, 7, Task 13

70. b. It needs to be robust

The project management plan is the document used to define the project work. It therefore needs to be robust enough to respond to changes in the project environment, which probably will occur in this situation since the project is to last three years. This agility may result in more information as the project progresses, and the project management plan is not baselined immediately as updates to this plan may be required many times. [Planning]

PMI®, *PMBOK® Guide*, 2017, 83
PMI® *PMP Examination Content Outline*, 2015, Planning, 7, Task 11

71. d. Activity name

Activity attributes are an output of the Define Activity process. The components for each activity evolve over time. In the initial stages of the project, they include the activity ID, WBS ID, and the activity name. Later, when completed, they may include activity codes, description, predecessor and successor activities, logical relationships, leads and lags, resource requirements, imposed dates, constraints, and assumptions. [Planning]

PMI®, *PMBOK® Guide*, 2017, 186
PMI® *PMP Examination Content Outline*, 2015, Planning, 6, Task 4

72. b. An external dependency

Some dependencies are external ones, and they involve a relationship between project activities and non-project activities. In sequencing activities, the project management team must determine which dependencies are external as they are usually outside of the team's control. Other examples of dependencies, a tool and technique in Sequence Activities, are mandatory, discretionary, and internal. [Planning]

PMI®, *PMBOK® Guide*, 2017, 191–192
PMI® *PMP Examination Content Outline*, 2015, Planning, 6, Task 4

73. d. Trend analysis

    Trend analysis is used in many control processes in project management. As a tool and technique in the data analysis category in Control Schedule, trend analysis examines the performance of the project over time to determine whether performance is improving or deteriorating. Graphical analysis techniques are valuable in trend analysis to understand performance to date and to compare it to future performance goals in the form of completion dates. [Monitoring and Controlling]

    PMI®, *PMBOK® Guide*, 2017, 227
    PMI® *PMP Examination Content Outline*, 2015, Monitoring and Controlling, 9, Task 1

74. c. Focusing on codified explicit knowledge

    In the Manage Project Knowledge process, it is a common misperception that the key to managing project knowledge is to only document it, so it can be shared. Another misperception is managing project knowledge consists solely of only codified, explicit knowledge can be shared in this way; however, codified explicit knowledge lacks context and then can easily lead to different interpretations as to its meaning and usefulness. While it can be shared, it may not be understood or applied as desired. Tacit knowledge, on the other hand, has context built into it, but it is difficult to codify. This means it resides in the heads of the people involved and is then shared through conversations and interactions between them. Surveys, if well-constructed by the knowledge management specialist and the people involved, and one-on-one discussions between the knowledge management specialist and the people involved are other methods to use. [Executing]

    PMI®, *PMBOK® Guide*, 2017, 100
    PMI® *PMP Examination Content Outline*, 2015, Executing, 8, Task 6

75. b. Accommodating

    Accommodating or smoothing is the preferred conflict resolution approach in this situation. In it, areas of agreement are emphasized rather than areas of difference. By doing so, one's position is conceded to that of the needs of the other person in this situation with the goal of maintaining harmony and relationships, the key words in this question. [Executing]

    PMI®, *PMBOK® Guide*, 2017, 349
    PMI® *PMP Examination Content Outline*, 2015, Executing, 8, Task 2

76. b. Lessons learned register

   In closing the project, the first output is project document updates. They may be marked as final versions since the project is closed. The lessons learned register is the one of particular importance since it is then finalized and includes information on the project, as in this example, or phase closure. This final lesson learned register may include information on benefits management, accuracy of the business case, project and development life cycles, risk and issue management, and stakeholder engagement. [Closing]

   PMI®, *PMBOK® Guide*, 2017, 127
   PMI® *PMP Examination Content Outline*, 2015, Closing, 10, Task 1

77. c. Can reduce construction claims

   There are numerous advances in tools in procurement management. Online tools are more prevalent and helpful for the buyer and the seller. In the construction and infrastructure field, the building information model is a software tool that has proven to save significant amounts of time and money. It enables a substantial reduction in construction claims, which then reduces the schedule and budget set aside for claims. Many organizations and governments throughout the world are mandating its use. [Monitoring and Controlling]

   PMI®, *PMBOK® Guide*, 2017, 463
   PMI® *PMP Examination Content Outline*, 2015, Monitoring and Controlling, 9, Task 7

78. b. Project charter

   Although the project charter cannot stop conflicts from arising, it can provide a framework to help resolve them, because it describes the project manager's authority to apply organizational resources to project activities. It is prepared by the project sponsor, as in this situation, and often in collaboration with the project manager as hopefully the project manager has been assigned and can participate. Regardless it is developed prior to the start of project planning. [Initiating]

   PMI®, *PMBOK® Guide*, 2017, 77, 81
   PMI® *PMP Examination Content Outline*, 2015, Initiating, 4, Task 5

79. a. Detect and control any defects or issues before you finish the project

There are five levels of increasingly effective quality management. The answer is the second one as you want to detect and control defects before deliverables are given to the customer. This is done as part of Control Quality. The Control Quality process has related costs, which primarily are appraisal costs and internal failure costs. [Monitoring and Controlling]

PMI®, *PMBOK® Guide*, 2017, 275
PMI® *PMP Examination Content Outline*, 2015, Monitoring and Controlling, 9, Task 3

80. c. Organizational change management

Organizational change management can influence team behavior. Change is not only limited to changes at the project level but also to changes at the organizational level. Rather than being concerned about solely managing changes to scope, time, cost, and quality, the project manager must embrace a broader approach and be aware of changes at the organizational level. These changes are strategic and may mean a change in the project's priority because of a new strategic objective that makes it less important; mergers or acquisitions such as the new company may have a similar project under way or recently completed; and the constant need to do more with less meaning the need to cut the budget or the number of resources, to list a few. The project manager needs to broaden his or her sphere of influence and work closely with members of the portfolio group or the strategic planning group to learn about any possible changes and then how to exploit and adapt them to benefit the project rather than taking a negative view. [Executing]

PMI®, *PMBOK® Guide*, 2017, 309
PMI® *PMP Examination Content Outline*, 2015, Executing, 8, Task 2

81. c. Create value for your team and for the suppliers

Quality is everyone's responsibility including suppliers. The work done by suppliers must be of the highest possible quality. Your goal to achieving success in this area is to establish a mutually beneficial relationship with the suppliers ideally with long-term relationships. This mutually beneficial relationship then reduces your work to Control Quality and is an emerging practice in quality management. It enhances the ability for your organization and that of the suppliers' organizations to create value for each other, enhances joint responses to customer needs and expectations, and optimizes costs and resources. [Monitoring and Controlling]

PMI®, *PMBOK® Guide*, 2017, 275
PMI® *PMP Examination Content Outline*, 2015, Monitoring and Controlling, 9, Task 3

82. d. Use multi-criteria decision analysis

    As a tool and technique in Acquire Resources, this approach enables
    criteria to be able to be developed and used to rate or score potential
    team members. The criteria are weighted given the team's requirements
    including items such as availability, cost, experience, ability, knowledge,
    skills, attitude, and international factors. [Executing]

    PMI®, *PMBOK® Guide*, 2017, 332
    PMI® *PMP Examination Content Outline,* 2015, Executing, 8, Task 2

83. d. A defined integrated change control process

    During the Control Resources process, change requests are an output.
    If changes occur in this process or when recommended corrective or
    preventive actions impact components of the project management plan,
    the project manager issues a change request. It is processed as defined
    in the Perform Integrated Change Control process. [Monitoring and
    Controlling]

    PMI®, *PMBOK® Guide*, 2017, 306
    PMI® *PMP Examination Content Outline,* 2015, Monitoring and
    Controlling, 9, Task 3

84. b. Use questionnaires and surveys

    In Control Quality, questionnaires and surveys are a data gathering
    technique. They are used to obtain data about customer satisfaction
    after the product or service is finished. Costs identified in these
    questionnaires and surveys may be considered external failure costs
    in the cost of quality, and If they occur, they then can have cost
    implications for the organization [Monitoring and Controlling]

    PMI®, *PMBOK® Guide*, 2017, 303
    PMI® *PMP Examination Content Outline,* 2015, Monitoring and
    Controlling, 9, Task 3

85. a. Provide special consideration to external policies

    Interpersonal and team skills are a tool and technique in the Acquire
    Resources process. They are primarily focused on negotiation as the
    project manager must negotiate with functional managers for resources.
    In this situation, part of your team will be located in a new country
    to your company. Given that no one has worked there before, special
    consideration is given to external negotiating policies, practices,
    processes, guidelines, and legal criteria. [Executing]

    PMI®, *PMBOK® Guide*, 2017, 332
    PMI® *PMP Examination Content Outline,* 2015, Executing, 8, Task 2

86. d. Holding performance reviews

   Performance reviews are a data analysis tool and technique in Control Schedule. Their purpose is to measure, compare, and analyze schedule performance against the schedule baseline. They may review actual start and finish dates, percent complete, and remaining schedule duration for work that is under way. [Monitoring and Controlling]

   PMI®, *PMBOK® Guide*, 2017, 227
   PMI® *PMP Examination Content Outline*, 2015, Monitoring and Controlling, 9, Task 1

87. d. Defect repair

   During the Direct and Manage Project Work process, work performance data are collected, actioned, and communicated. This process requires review of the impact of project changes and the implementation of proposed changes, one of which is defect repair. It is an intentional activity to modify a nonconforming product or project component. Work performance information is an output of this process; other examples are work completed, KPIs, technical performance resources, actual start and finish dates of schedule activities, story point compiled, deliverable status, schedule, process, number of change requests, actual costs incurred, and actual durations. [Executing]

   PMI®, *PMBOK® Guide*, 2017, 95
   PMI® *PMP Examination Content Outline*, 2015, Executing, 7, Task 1

88. c. Someone external to the project

   The charter is issued by an entity external to the project, such as a sponsor, a program manager, a PMO, or a portfolio management chair or an authorized representative. The initiator or sponsor should be at a level such that he or she can provide funding and commit to the project. [Initiating]

   PMI®, *PMBOK® Guide*, 2017, 77
   PMI® *PMP Examination Content Outline*, 2015, Initiating, 5, Task 6

89. c. Invite your customer and his team to project meetings

An emerging trend in communications management is to include stakeholders in project meetings. In this approach, external stakeholders, such as the customer, are invited as well as other stakeholders inside the organization. It is also a practice common in agile processes in which during the daily stand-up meetings, any achievements and issues from the previous day are discussed, and the team discusses plans for the current day. Customer involvement may be useful in resolving issues that have occurred. [Executing]

PMBOK® Guide, 2017, 364
PMI® *PMP Examination Content Outline*, 2015, Executing, 8, Task 6

90. d. Identify and deal with resource shortages

The Control Resources process is concerned with physical resources and is performed throughout the project. Updating resource allocation requires knowing the actual resources that have been used, and the resources that still are needed. The project manager reviews performance to date. He or she then identifies resource shortages in a timely way. Since much of the physical resources are acquired through contracts/agreements, including leasing, the project manager also tracks the resources that are no longer required to close the contract or end the lease. [Monitoring and Controlling]

PMI®, *PMBOK® Guide*, 2017, 355
PMI® *PMP Examination Content Outline*, 2015, Monitoring and Controlling, 9, Task 1

91. c. Resource configurations

Controlling physical resources involves allocating and using material, equipment, and supplies needed for the project for its successful completion. Organizations that use a large amount of physical resources should have data on resources needed now and in the future. They also should have resource configurations, the answer to this question, that will be required to meet the demands and the supply of the resources. Successful project manager are ones who can manage and control needed physical resource effectively; otherwise, it is a source of risk to the project. [Monitoring and Controlling]

PMI®, *PMBOK® Guide*, 2017, 310
PMI® *PMP Examination Content Outline*, 2015, Monitoring and Controlling, 9, Task 1

92. b. You need to reduce activity costs as much as possible

This question shows an opportunity has been recognized, which means the risk register is being used as it is an input in project documents to the Estimate Costs process. It should be reviewed to as it contains details of project risks that have been identified and prioritized and need risk responses, which tend to involve costs. It provides detailed information that can be used to estimate costs. Risks, whether threats or opportunities, have an impact on both activity and overall project costs. In this example, it is a potential opportunity that can benefit the business depending on the response action selected as it may reduce directly activity costs or accelerate the schedule. [Planning]

PMI®, *PMBOK® Guide*, 2017, 242
PMI® *PMP Examination Content Outline*, 2015, Planning, 6, Task 3

93. a. Cost management plan

The management and control of costs focuses on variances. The cost management plan includes control thresholds, also called variance thresholds. Certain variances are acceptable, and others, usually those falling outside a particular range, are unacceptable. The actions taken by the project manager for all variances are described in the cost management plan. These thresholds are typically expressed as percentage deviations from the baseline plan. [Planning]

PMI®, *PMBOK® Guide*, 2017, 239
PMI® *PMP Examination Content Outline*, 2015, Planning, 6, Task 3

94. b. Norming

There are four stages of team development: forming, storming, norming, and performing in the original team development model developed by Tuckman; later adjourning was added as a fifth stage. In the norming stage team members begin to work together, adjusting their work habits and behaviors to support the team. The team members learn to trust each other. [Executing]

PMI®, *PMBOK® Guide*, 2017, 338
PMI® *PMP Examination Content Outline*, 2015, Executing, 8, Task 2

95. a. Manage knowledge transfer

    As you close your project, you cannot overlook the importance of knowledge transfer. Each person on the project brings knowledge assets and acquires others as he or she works on the project. By managing knowledge transfer, you are focusing on sharing knowledge for other projects under way or to be done in the future. It differs from collecting lessons learned, also important in closing, as its focus is on knowledge transfer and sharing knowledge, not hoarding it. [Closing]

    PMI®, *PMBOK® Guide*, 2017, 123
    PMI® *PMP Examination Content Outline,* 2015, Closing, 10, Task 5

96. b. Informal scope control related practices are used

    In this situation you are wondering whether you need to tailor some approaches in your project management methodology as they may not be needed. Tailoring considerations in Scope Management need to consider several areas, one of which is validation and control. In this area, the organization may have existing formal or informal validation and control related policies, procedures, and guidelines so tailoring is not needed. [Monitoring and Controlling]

    PMI®, *PMBOK® Guide*, 2017, 133
    PMI® *PMP Examination Content Outline,* 2015, Monitoring and Controlling, 9, Task 1

97. c. Iterative scheduling with a backlog

    Iterative scheduling with a backlog is an emerging trend in developing a schedule. It is a form of rolling wave planning based on an adaptive life cycle and an agile approach in product development. Requirements are documented in user stories and are prioritized and refined before construction. Product features use time-based periods of work. Multiple teams can work concurrently to develop a large number of features with few interconnected dependencies. This approach also is open to changes throughout the development life cycle. [Planning]

    PMI®, *PMBOK® Guide*, 2017, 177
    PMI® *PMP Examination Content Outline,* 2015, Planning, 5, Task 4

98. d. Engaging stakeholders at certain stages

   While the other answers are good practices, this answer is part of the Manage Stakeholder Engagement process. It shows the necessity of engaging stakeholders at appropriate project stages to obtain, confirm, and maintain their continual commitment to the project. If a key stakeholder is missing meetings, it is critical for the project manager to meet with this stakeholder and find out his or her concerns and actively listen. If the stakeholder has concerns, then the project manager should address them and see if he or she can regain the stakeholder's support. [Executing]

   PMI®, *PMBOK® Guide*, 2017, 524
   PMI® *PMP Examination Content Outline*, 2015, Executing, 8, Task 7

99. b. Hold some meetings

   In the Monitor Stakeholder Engagement process, meetings are a tool and technique. They include status meetings, standup meetings retrospectives, and any others listed in the stakeholder engagement plan. These meetings are held to monitor and assess stakeholder engagement levels. They can be face to face or virtual. [Monitoring and Controlling]

   PMI®, *PMBOK® Guide*, 2017, 535
   PMI® *PMP Examination Content Outline*, 2015, Monitoring and Controlling, 8, Task 1

100. d. Developing a communications strategy

   A communications strategy is the first step. It is based on the project's needs and those of the stakeholders. Once the strategy is determined, then the communications management plan is prepared, which among other things ensures stakeholders receive messages in various formats and means as defined by the strategy. These messages constitute the planning process, which is the second part for successful communications. Using the communications strategy and the communications management plan then provide the foundation to monitor the effect of communications and hopefully avoid misunderstandings and mis-communications. [Planning]

   PMI®, *PMBOK® Guide*, 2017, 362
   PMI® *PMP Examination Content Outline*, 2015, Planning, 6, Task 6

101. a. Alternatives analysis

In Plan Schedule Management, a tool and technique to use is data analysis, which includes alternatives analysis. It can include the scheduling method to use or how to combine different scheduling methods. It also can include the level of detail for the schedule, the need for rolling wave planning, and how often the schedule should be reviewed and updated. Consider a balance between the level of detail and the amount of time it will take to update it on the project. [Planning]

PMI®, *PMBOK® Guide*, 2017, 181
PMI® *PMP Examination Content Outline*, 2015, Planning, 6, Task 4

102. d. Should be consistent with the portfolio management plan

Therefore, if the portfolio management plan requires review by the key stakeholders each time it is revised, the same is the case for when the project management plan is revised. [Planning]

PMI®, *PMBOK® Guide*, 2017, 83
PMI® *PMP Examination Content Outline*, 2015, Planning, 7, Task 11

103. b. Resources for communications activities

Of the answers to this question, resources are included in the communications management plan. Resources need to be allocated for communications activities. Without resources, these critical activities would not be performed. Also, in this plan while resources include human resources, they also include time and budget considerations. [Planning]

PMI®, *PMBOK® Guide*, 2017, 377
PMI® *PMP Examination Content Outline*, 2015, Planning, 6, Task 6

104. d. Focus on goals to be served

Decision making involves the ability to negotiate and influence the organization and the project management team. This answer is the first guideline; the other answers refer to items in leadership and influencing; all are examples of interpersonal skills of the project manager, a tool and technique in the Manage Team process. [Executing]

PMI®, *PMBOK® Guide*, 2017, 349
PMI® *PMP Examination Content Outline*, 2015, Executing, 8, Task 2

105. b. Have meetings

All four answers are tools and techniques in the Develop Project Charter process. However, in this situation, the best answer is to conduct meetings with the functional department leaders since you need resources from these groups for project success. These meetings are useful to discuss the project objectives, success criteria, key deliverables, high-level requirements, and other information. The meetings enable you to have an understanding of any concerns these managers may have and listen to them and provide feedback to enlist their support. [Initiating]

PMI®, *PMBOK® Guide*, 2017, 80
PMI® *PMP Examination Content Outline*, 2015, Initiating, 5, Task 5

106. d. Accelerate knowledge sharing

In an agile environment to Monitor Risk, adaptive approaches are used through frequent reviews of incremental work products. It also is necessary to ensure the cross-functional team accelerates knowledge sharing and ensures risks are understood and managed. Requirements are updated regularly in order that risks can be reprioritized as the project progresses. [Monitoring and Controlling]

PMI®, *PMBOK® Guide*, 2017, 400
PMI® *PMP Examination Content Outline*, 2015, Monitoring and Controlling, 9, Task 4

107. c. Organizational strategy

Expert judgment is a tool and technique in the Develop Project Charter process. It is defined as judgment provided based on experience in an application area, knowledge area, discipline, or industry based on the specific project. It may be provided by a group or by an individual. In developing the charter, experts are useful in organizational strategy, benefit management, technical knowledge of the project's industry and area of focus, duration and budget, and risk identification. Organizational strategy is useful since one purpose of the charter is to link the project to the organization's strategic objectives. [Initiating]

PMI®, *PMBOK® Guide*, 2017, 75, 79
PMI® *PMP Examination Content Outline*, 2015, Initiating, 5, Task 7

108. b. Reporting

Meetings are a tool and technique in the Close Project or Phase process. They are used to confirm deliverables are accepted, to validate the exit criteria to realize contact completion, to evaluate stakeholder satisfaction, to collect lessons learned, to transfer knowledge and information, and to celebrate success. Close-out reporting meetings are an example of a type of meeting that may be held. [Closing]

PMI®, *PMBOK® Guide*, 2017, 127
PMI® *PMP Examination Content Outline*, 2015, Closing, 10, Task 7

109. b. Mind mapping

In Plan Stakeholder Engagement, mind mapping is a data representation tool and technique to consider. It is useful because it visually organizes information about stakeholders, and it also shows their relationships to one another. [Planning]

PMI®, *PMBOK® Guide*, 2017, 521
PMI® *PMP Examination Content Outline*, 2015, Planning, 7, Task 13

110. c. Refer to the team charter

In this situation, since the team members are having conflicts and are stakeholders to engage, you need to meet with the team and remind them of the team charter, which each team member agreed to follow and signed. The team charter is the ground rules for the team, a tool and technique in this process. It contains the expected behavior for the team members with regard to stakeholder engagement. [Executing]

PMI®, *PMBOK® Guide*, 2017, 319–320, 528
PMI® *PMP Examination Content Outline*, 2015, Executing, 8, Task 7

111. b. Technical documentation

Procurement documentation requires review before contracts are closed and later before the project is closed. The procurement documents are collected, indexed, and filed. It is easy to focus on cost/schedule information as an example, but technical documentation should not be overlooked in the process. Such information includes 'as-built" plans/drawings, "as-developed" documents, manuals, and troubleshooting as this information can be used for lessons learned and to evaluate contractors for future contracts. [Closing]

PMI®, *PMBOK® Guide*, 2017, 125
PMI® *PMP Examination Content Outline*, 2015, Closing, 10, Task 3

112. c. Transfer your work to human resources

In the Close Process or Phase process, administrative closure procedures include actions and activities to transfer the project's products, services, or results (as in this example) to the next phase or to production or operations as in this example. Since the project involves mentoring, the human resource department or unit is the part of the organization to continue this work. [Closing]

PMI®, *PMBOK® Guide*, 2017, 101
PMI® *PMP Examination Content Outline*, 2015, Closing, 10, Task 2

113. b. Release resources when the project is in the closing phase

The closing process is when all activities across all process groups are complete. It verifies the project then project work is done, and the project has met its objectives. While a variety of activities are performed, one is to and release resources. The other answers are nice practices to follow. [Closing]

PMI®, *PMBOK® Guide*, 2017, 123
PMI® *PMP Examination Content Outline*, 2015, Closing, 10, Task 2

114. c. Assumption log

The other possible answers are part of the project charter. The assumption log is the other output of this process. While high-level assumptions and constraints are typically in the business case, lower level activity and task assumptions are generated during the project. They are then placed in this assumption log and used throughout the project. [Initiating]

PMI®, *PMBOK® Guide*, 2017, 81
PMI® *PMP Examination Content Outline*, 2015, Initiating, 5, Task 4

115. b. Emphasize collaboration

In an agile/adaptive environment, collaboration is necessary to boost productivity and facilitate problem solving. A collaborative team is one that may facilitate accelerated integration of distinct work activities, improve communications, increase knowledge transfer, and promote flexibility in work assignments. [Executing]

PMI®, *PMBOK® Guide*, 2017, 311
PMI® *PMP Examination Content Outline*, 2015, Executing, 8, Task 2

116. b. Involving team members in planning is beneficial.

    The project manager is the leader and manager of the team. Often, resources are preassigned to the project or have been identified in the charter. Although roles and responsibilities will be assigned, involving team members in planning and decision making is beneficial. Their participation when possible adds their expertise to the project, and most importantly, it helps strengthen their commitment to the project objectives. [Planning]

    PMBOK® Guide, 2017, 309
    PMI® *PMP Examination Content Outline*, 2015, Planning, 6, Task 5

117. a. Use it to build a collaborative working environment

    Team building consists of activities to enhance the team's social relations and build a collaborative and cooperative working environment. Team building can be part of any meeting even as a five-minute agenda item. The team-building objective is to help individuals work together effectively as a team. It is ongoing for project success. [Executing]

    PMI®, *PMBOK® Guide*, 2017, 341
    PMI® *PMP Examination Content Outline*, 2015, Executing, 8, Task 2

118. c. Influencing

    All are useful skills for project managers. However, in the Control Resources process, negotiation and influencing are the two interpersonal skills tools and techniques to consider. In this situation influencing was necessary as the project manager has little or no direct control over team members as they work in a weak matrix. The ability of the project manager to influence stakeholders in a timely basis is critical to project success. By using influencing, the project manager can solve problems and obtain resources needed in a timely way. [Monitoring and Controlling]

    PMI®, *PMBOK® Guide*, 2017, 357
    PMI® *PMP Examination Content Outline*, 2015, Monitoring and Controlling, 9, Task 1

119. a. Periodically review the business case

    The business case is not a one-time document but is used throughout the project, regardless of whether the project has multiple phases. It helps measure the project's success against the project outcomes. It is an input to the Develop Project Charter process since it describes the information from a business standpoint to determine whether the expected outcomes from the project are worth the required investment. [Initiating]

    PMI®, *PMBOK® Guide*, 2017, 30, 77
    PMI® *PMP Examination Content Outline*, 2015, Initiating, 5, Task 2

120. a. Conduct a risk audit

    The risk audit is a tool and technique in Monitor Risks to evaluate the effectiveness of the risk management process. The project manager is responsible for ensuring these audits are done at the frequency defined in the project risk management plan. They may be part of the regular project review meeting, may be part of a risk review, or there may be separate meetings for the risk audit. [Monitoring and Controlling]

    PMI®, *PMBOK® Guide*, 2017, 456
    PMI® *PMP Examination Content Outline*, 2015, Monitoring and Controlling, 9, Task 4

121. c. Conflict management skills are useful

    Interpersonal and team skills are a tool and technique in the Develop Project Charter process; one of them is conflict management skills. Even if your charter provides you with the authority to acquire the resources you need, unless you work in a pure projectized structure, it still is a challenge to get the resources you need when you need them. Functional managers will be reluctant to give you the key people you require. These conflict management skills can bring stakeholders into alignment on various aspects in the charter. [Initiating]

    PMI®, *PMBOK® Guide*, 2017, 80
    PMI® *PMP Examination Content Outline*, 2015, Initiating, 5, Task 5

122. c. Technical performance analysis

    Technical performance analysis is a data analysis tool and technique in Monitor Risks. Its purpose is compare technical performance to the schedule of technical achievement. It requires the definition of objective, quantifiable measures of technical performance to help compare actual results against targets. Example of these measures are weight, transaction times, number of delivered defects, or storage capacity. Deviation can help indicate the possible impact of threats or opportunities. [Monitoring and Controlling]

    PMI®, *PMBOK® Guide*, 2017, 456
    PMI® *PMP Examination Content Outline*, 2015, Monitoring and Controlling, 9, Task 4

123. c. Hold a meeting

    Meetings are a tool and technique in Monitor Risks. These meetings can include risk reviews. They are scheduled regularly to evaluate the effectiveness of risk responses in dealing with both overall and individual risks. These reviews may identify new risks, identify secondary risks, reassess existing risks, close outdated risks, close issues from risks that have occurred, and identify lessons learned. [Monitoring and Controlling]

    PMI®, *PMBOK® Guide*, 2017, 457
    PMI® *PMP Examination Content Outline*, 2015, Monitoring and Controlling, 9, Task 4

124. b. You may accept more risks

    In procurement management an emerging trend is greater attention to risk management. No contractor can manage all the risks on a project. As you work in the Control Procurements process, you will be required to accept the risks in which the contractor lacks control, such as changing corporate policies, changing legislative or regulatory requirements, or other changes external to the project. [Monitoring and Controlling]

    PMI®, *PMBOK® Guide*, 2017, 463
    PMI® *PMP Examination Content Outline*, 2015, Monitoring and Controlling, 9, Task 7

125. a. Procurement performance review

    These reviews are a tool and technique in the Control Procurements process under data analysis, which includes measure, control, and analyze quality, resources, schedule, and cost performance against that in the agreement. These reviews also include identifying work packages that already are behind schedule, over or under budget, or have resource or quality issues. Inspections and audits, answers b and c, are specified in the contract and are required by the buyer to verify compliance in the seller's work processes or deliverables; they also are a tool and technique in this process. A status review meeting, answer d, is an informal session to discuss performance to date. [Monitoring and Controlling]

    PMI®, *PMBOK® Guide*, 2017, 498
    PMI® *PMP Examination Content Outline*, 2015, Monitoring and Controlling, 9, Task 7

126. b. Warranty work

Warranty work is an example of the cost of nonconformance. These costs are divided into internal failure costs and external failure costs. They are money spent after the project because of project failures. Other external failure costs are liabilities and lost business. [Planning]

PMI®, *PMBOK® Guide*, 2017, 283
PMI® *PMP Examination Content Outline*, 2015, Planning, 6, Task 8

127. a. Workmanship standards

Workmanship standards are an example of an enterprise environmental factor in addition to government or industry standards, quality standards, and safety standards. The other answers are examples of organizational process assets, another input to the Develop Project Charter process. [Initiating]

PMI®, *PMBOK® Guide*, 2017, 78–79
PMI® *PMP Examination Content Outline*, 2015, Initiating, 5, Task 1

128. b. Diversity

Many teams are comprised with diverse team members from different backgrounds and cultures. A diverse team also is useful in working together to share ideas as diversity can foster collaboration and problem solving by merging different views into a more effective solution. It is noted as a reason to tailor how resource management is used on your team. [Executing]

PMI®, *PMBOK® Guide*, 2017, 317
PMI® *PMP Examination Content Outline*, 2015, Executing, 8, Task 2

129. d. Direct the right level of communications to the stakeholders

In Plan Stakeholder Engagement, the stakeholder engagement assessment matrix is a data representation tool and technique. As in Identify Stakeholders, it shows the current engagement levels of each stakeholder as compared to the desired levels for successful project delivery. In this process, it is used to identify gaps and to direct the appropriate level of communication needed to engage each stakeholder. [Planning]

PMI®, *PMBOK® Guide*, 2017, 521–522
PMI® *PMP Examination Content Outline*, 2015, Planning, 7, Task 6

130. c. Use a stakeholder cube

    The stakeholder cube is a data representation tool and technique in Identify Stakeholders. It refines the three grids and combines the grid elements into a three-dimensional model. The project manager and team then use this grid to identify and engage the stakeholder community. The model has multiple dimensions to improve how the stakeholder community is shown as it is presented in a multidimensional way and can help develop communications strategies. [Initiating]

    PMI®, *PMBOK® Guide*, 2017, 513
    PMI® *PMP Examination Content Outline*, 2015, Initiating, 5, Task 3

131. d. Direct and Manage Project Work

    Work performance data containing these examples are an output of Direct and Manage Project Work. They are raw observations and measurements identified as activities are being performed to complete the work of the project. These data often are viewed at the lowest level of detail from which information is derived by other processes. The data then are gathered as the work is done and passed to the controlling processes of the various processes for further analyses. [Executing]

    PMI®, *PMBOK® Guide*, 2017, 95
    PMI® *PMP Examination Content Outline*, 2015, Executing, 8, Task 6

132. d. Forcing

    Forcing, or directing, using power or dominance, implies the use of position power to resolve conflict. It involves imposing one viewpoint at the expense of another. Project managers may use it when time is of the essence, when an issue is vital to the project's well-being, or when they think they are right based on available information. It offers only win-lose situations. Although this approach is appropriate when quick decisions are required or when unpopular issues are an essential part of the project, it puts project managers at risk. [Executing]

    PMI®, *PMBOK® Guide*, 2017, 349
    PMI® *PMP Examination Content Outline*, 2015, Executing, 8, Task 2

133. b. The project charter

The project charter is an input in developing the project management plan. It is the starting point for initial planning and at a minimum, the charter defines the high-level information about the project to be expanded upon in the various parts of the project management plan. Outputs from other processes, organizational process assets, and enterprise environmental factors are other inputs to consider. [Planning]

PMI®, *PMBOK® Guide*, 2017, 83
PMI® *PMP Examination Content Outline*, 2015, Planning, 7, Task 11

134. c. Issue a change request

Change requests are an output of the Manage Project Team process. Staffing changes can affect the rest of the team, even if by choice in this situation or by an uncontrollable event. When staffing changes affect the project, such as causing the schedule to slip or the budget to be exceeded, a change request should be processed through the Perform Integrated Change Control process. While in this question the change involved moving the team member to a different assignment, staffing changes also involve outsourcing the work and replacing team members who leave. [Executing]

PMI®, *PMBOK® Guide*, 2017, 350
PMI® *PMP Examination Content Outline*, 2015, Executing, 8, Task 2

135. d. Identify all potential project stakeholders and their relevant information

Stakeholder analysis is data analysis tool and technique in Identify Stakeholders. Its purpose is to identify a list of stakeholders and relevant information about each one. The information may include the position in the organization, project roles, stakes in the project, expectations, and attitudes. Stakes may be whether the stakeholders are interested, have legal or moral rights, have ownership, have specialized ownership, or are a contributor. [Initiating]

PMI®, *PMBOK® Guide*, 2017, 512
PMI® *PMP Examination Content Outline*, 2015, Initiating, 5, Task 3

136. b. Project management plan

Project scope is measured against the project management plan. The project scope statement and scope baseline are subsets of the project management plan. However, the whole plan and all the baselines (cost and schedule) need to be met in addition to scope. The project management plan is the agreement between the project manager and sponsor and defines what constitutes project completion. [Closing]

PMI®, *PMBOK® Guide*, 2017, 123
PMI® *PMP Examination Content Outline*, 2015, Closing, 10, Task 1

137. c. Follow an adaptive approach

Large projects may use an adaptive approach for some deliverables, while for others a more stable approach may be used. Given the size and the geographic locations involved, this situation is one in which an adaptive approach is more suitable. A governing agreement such as a master services agreement may be used for the project. The adaptive approach may be part of an appendix. It shows changes to occur on adaptive scope without impacting the overall agreement, useful in Control Procurement. [Planning]

PMI®, *PMBOK® Guide*, 2017, 465
PMI® *PMP Examination Content Outline*, 2015, monitoring and Controlling, 9, Task 8

138. c. Assess the schedule objectives to see if benefits were achieved

The other answers are variations of the tools and techniques used in Close Project or Phase. An output is the final report, which provides a summary of project performance. While it contains many items, it evaluates the schedule objectives to see if results achieved the benefits the project was undertaken to address. If the benefits were not met when the project was closed, then this section indicates the degree to which they were achieved and estimates what was needed for future benefit realization. [Closing]

PMI®, *PMBOK® Guide*, 2017, 127–128
PMI® *PMP Examination Content Outline*, 2015, Closing, 10, Task 4

139. b. Identify the causes of the conflict

Conflicts are inevitable on projects. They may include scarce resources, scheduling practices, and personal work styles. However, successful conflict resolution can result in increased productivity and positive working relationships. If managed properly, these differences may lead to enhanced creativity and more effective decision making. In order to address conflict, people must recognize and acknowledge that conflict exists. The project manager must be able to identify the causes of the conflict and then actively manage it to avoid any negative impacts. Conflicts should be addressed early and usually in private with a direct, collaborative approach. [Executing]

PMI®, *PMBOK® Guide*, 2017, 348
PMI® *PMP Examination Content Outline*, 2015, Executing, 8, Task 2

140. c. Determine the number of languages used

It is often necessary in communications management to tailor how processes are used. In Monitor Communications, language is a key consideration. It is easy to consider English as the language preferred by the team, but with the increase in virtual teams and teams with people from different cultures, it may not be the preferred language to use. However, it is necessary to determine how many languages are used and then determine whether any allowances have been made to adjust to team members from different language groups. One approach to consider if diverse languages are spoken by team members as their primary language is to adopt English but focus on the key 1,000 words that commonly are used to ensure there are no misunderstandings based on slang or the use of words that are not used frequently. [Monitoring and Controlling]

PMI®, *PMBOK® Guide*, 2017, 365
PMI® *PMP Examination Content Outline*, 2015, Monitoring and Controlling, 9, Task 1

141. a. Decision making

Both processes use inspection. Validate Scope also uses decision making tools and techniques. Voting is a decision-making technique used to reach a conclusion when stakeholders and the project team perform the validation. [Monitoring and Controlling]

PMI®, *PMBOK® Guide*, 2017, 166
PMI® *PMP Examination Content Outline*, 2015, Monitoring and Controlling, 9, Task 3

142. c. Knowledge

In this situation, the key word is 'client'. In acquiring resources for your team, you can use multi-criteria decision analysis, a tool and technique in this process. The criteria are used to rate and score team members and can be weighted to show their importance. All of the answers to this question are important, but knowledge is the most critical since it considers whether the team member has relevant knowledge of the customer, focusing on this question. The knowledge criterion also involves similar related projects and nuances of the project environment. Other criteria not shown in the answers include availability, eliminated since it is a top project in the company, and one should then be able to acquire the resource; cost, eliminated for the same reason; attitude, which is desirable to be able to work well with the team; and international factors, eliminated since the question does not note it is a global project. Experience, ability, and skills are critical, but knowledge is selected as it emphasizes the customer. [Executing]

PMI®, *PMBOK® Guide*, 2017, 332
PMI® *PMP Examination Content Outline*, 2015, Executing, 8, Task 2

143. d. Forcing

Forcing or directing is represented by a strong desire to satisfy oneself rather than to satisfy others. It involves imposing one viewpoint at the expense of another and is characterized by a win-lose outcome in which one party overwhelms the other. It pushes one's view at the expense of others usually through a power position to resolve a conflict. [Executing]

PMI®, *PMBOK® Guide*, 2017, 349
PMI® *PMP Examination Content Outline*, 2015, Executing, 8, Task 2

144. b. Part of questionnaires and surveys

Questionnaires and surveys are a data gathering technique, a tool and technique in Identify Stakeholders. Also, in this category are focus groups, which are used to gauge the attitudes of stakeholders about the project and enables the project manager and team to learn their level of support for it and any issues they may have so these issues can be resolved quickly. [Initiating]

PMI®, *PMBOK® Guide*, 2017, 511
PMI® *PMP Examination Content Outline*, 2015, Initiating, 5, Task 3

145. a. The choice of media

The choice of media, or the way you deliver the information, is as important as what you say. It is important to determine when to communicate in writing versus orally, when to prepare an informal memo or when to use a formal report, and when to communicate face to face or by email, as examples. Other examples are when to use push or pull options and choosing the technology to use. [Executing]

PMI®, *PMBOK® Guide*, 2017, 381
PMI® *PMP Examination Content Outline*, 2015, Executing, 8, Task 6

146. c. Issue a change request

Change requests are an output of Implement Risk Responses. The key word in the question was cost. Implementing risk responses may require a change request to the cost and schedule baselines. These change requests then are processed for review and disposition using the Perform Integrated Change Control process. [Executing]

PMI®, *PMBOK® Guide*, 2017, 451
PMI® *PMP Examination Content Outline*, 2015, Executing, 8, Task 5

147. c. Stakeholder register

The stakeholder register is the main output of Identify Stakeholders and contains all the details known at the time related to the stakeholders including identification information, assessment information, and stakeholder classification. The project manager and team should review the stakeholder register regularly since stakeholder identification is ongoing throughout the project, and different stakeholders may have different interests, expectations, or potential influence than those initially identified. [Initiating]

PMI®, *PMBOK® Guide*, 2017, 514
PMI® *PMP Examination Content Outline*, 2015, Initiating, 5, Task 3

148. a. Library services

Of the answers listed, library services are the only tool and technique to create and connect people to information. These tools are useful to share simple, unambiguous, and codified explicit knowledge. These tools, however, can be enhanced by adding some type of information such as 'contact me', so people can contact the originators and ask for advice as needed. The other examples are ones to connect people to people to work together to create new knowledge, share explicit knowledge, and integrate the team member's diverse knowledge. [Executing]

PMI®, *PMBOK® Guide*, 2017, 103
PMI® *PMP Examination Content Outline*, 2015, Executing, 8, Task 6

149. d. Basis of estimates

Basis of estimates is an output from the Estimate Costs process. It provides details to support the cost estimate to provide a clear and understandable approach to how the costs were estimated. Typical contents include how the estimate was developed, assumptions and constraints, identified risks as costs were estimated, the range of possible estimates, and the level of confidence in the final cost estimate. [Planning]

PMI®, *PMBOK® Guide*, 2017, 247
PMI® *PMP Examination Content Outline*, 2015, Planning, 6, Task 3

150. a. Develop Project Charter

In the Develop Project Charter process, an agreement is an input to define initial intentions for the project. They may be known as an understanding or memorandum of understanding, a service level agreement, letter of intent, a verbal agreement, or an e-mail. Agreements are an input to the Develop Project Charter process. [Initiating]

PMI®, *PMBOK® Guide*, 2017, 78
PMI® *PMP Examination Content Outline*, 2015, Initiating, 5, Task 5

151. c. Accepted deliverables

Accepted deliverables are an input to the Close Project or Phase. They may include approved product specifications, delivery receipts, and work performance documents. Partial or interim deliverables may be included if it is a phased or a canceled project. Other inputs are the project charter, the project management plan, project documents, business documents, agreements, procurement documentation, and organizational process assets. Expert judgment is a tool and technique in this process. [Closing]

PMI®, *PMBOK® Guide*, 2017, 125
PMI® *PMP Examination Content Outline*, 2015, Closing, 10, Task 1

152. d. Use trend analysis

Data gathering is the tool and technique in Control Scope. It includes variance analysis and trend analysis. Trend analysis reviews project performance over time to see if performance is improving or deteriorating. Its focus is to determine the cause and amount of variance in relation to the scope baseline to be able to determine if preventive or corrective action is required. [Monitoring and Controlling]

PMI®, *PMBOK® Guide*, 2017, 170
PMI® *PMP Examination Content Outline*, 2015, Monitoring and Controlling, 9, Task 1

153. c. A high-level requirement

    The project charter formally authorizes the existence of the project and provides the project manager with the organizational resources for the project activities. Using agile is an example of a high-level requirement, which is one of the many items included in the charter. [Initiating]

    PMI®, *PMBOK® Guide*, 2017, 81
    PMI® *PMP Examination Content Outline*, 2015, Initiating, 5, Task 5

154. c. Note key stakeholders as parties in the contract

    Contracts are mutually binding agreements that obligate the seller to provide products services or results and obligates the buyer to compensate the seller. These contracts or agreements are legal relationships subject to remedy in the courts. Agreements are an input to the Identify Stakeholder process. If the project results from a procurement activity or is based on an established agreement/contract, the parties in the contract are key project stakeholders. Others, such as suppliers, are also stakeholders and should be added to the stakeholder list. [Initiating]

    PMI®, *PMBOK® Guide*, 2017, 489, 510
    PMI® *PMP Examination Content Outline*, 2015, Initiating, 5, Task 3

155. a. Use an impact/influence grid

    All four answers are data representation techniques and can help categorize your stakeholders to help build relationships with them. The key words in the question is it is a small project. Since it is small, a power/interest, power/influence, or impact/influence [the answer] grid are appropriate ways to categorize them. Each of these three grids categorizes them according to their level of authority, level of concern, ability to influence the project's outcomes, or the ability to cause changes to the project's planning or execution. These models are used for small projects or ones with relationships that are relatively simple between the stakeholders and the project or within the stakeholder community. [Initiating]

    PMI®, *PMBOK® Guide*, 2017, 512
    PMI® *PMP Examination Content Outline*, 2015, Initiating, 5, Task 3

156. b. Use an issue log

An issue log is an output of the Direct and Manage Project Work process. It lists items such as: issue type, who raised it and when, description, priority, person assigned to the issue, target resolution date, status, and the final solution. This issue log can help the project manager track and manage issues, investigate them, and resolve them; recognizing issues may occur at any time on the project. [Executing]

PMI®, *PMBOK® Guide*, 2017, 85
PMI® *PMP Examination Content Outline*, 2015, Executing, 8, Task 1

157. a. Role

The resource plan documents roles, responsibilities, authority, and competence on the project. A role is the function assumed by or assigned to a person in the project. The court liaison in this question is an example of such a role on a project. Other examples are civil engineer, business analyst, or testing coordinator. Role clarity concerning authority and responsibilities also should be documented to help avoid negative conflicts later in the project. [Planning]

PMI®, *PMBOK® Guide*, 2017, 318–319
PMI® *PMP Examination Content Outline*, 2015, Planning, 6, Task 5

158. b. Determine how and when project benefits will be delivered

In Identify Stakeholders in the first iteration of doing so, business documents should be reviewed, one of which is the benefits management plan. The purpose of this plan is to realize the benefits in the project's business case. It may include the people who will benefit the most when the outcomes of the project are complete as they are stakeholders and beneficiaries. This plan includes a description of the target benefits, both tangible and intangible, the stakeholders will realize, how and when the benefits will be delivered, and how the benefits will be measured. [Initiating]

PMBOK® Guide, 2017, 33, 509
PMI® *PMP Examination Content Outline*, 2015, Initiating, 5, Task 7

159. b. Review your project charter

The project charter should be reviewed first as you identify stakeholders since it includes an identified stakeholder list. The charter also may contain information on stakeholder responsibilities. [Initiating]

PMI®, *PMBOK® Guide*, 2017, 395
PMI® *PMP Examination Content Outline*, 2015, Initiating, 5, Task 8

160. c. Understand as much as possible about the communications receiver

There are three fundamental activities of effective communications and developing effective communications artifacts. One is the answer to the question. One should understand as much as possible about the receiver, which includes needs and preferences, rather than relying on the traditional sender-receiver model. It also is necessary to have clarity as to the purpose of the communication rather than the push communications approach, and then to monitor and measure whether the communications were effective. [Planning]

PMI®, *PMBOK® Guide*, 2017, 363
PMI® *PMP Examination Content Outline*, 2015, Planning, 6, Task 6

161. b. Management is responsible for providing needed resources

An emerging practice in project quality management is management responsibility. While the quality management plan states resource roles and responsibilities, management retains in its requirements for quality the responsibility to provide suitable resources with adequate capacities. It also should be noted that quality success requires the participation of all project team members. [Planning]

PMI®, *PMBOK® Guide*, 2017, 275
PMI® *PMP Examination Content Outline*, 2015, Planning, 6, Task 8

162. d. Enterprise environmental factor

Enterprise environmental factors and organizational process assets are both inputs to Plan Quality Management. In this situation, the team is located in different geographical areas, and geographic distribution is an enterprise environmental factor to consider. Another one, which fits this question, is the organizational structure. Enterprise environmental factors are typically outside of the project manager's responsibilities. [Planning]

PMI®, *PMBOK® Guide*, 2017, 280
PMI® *PMP Examination Content Outline*, 2015, Planning, 6, Task 8

163. d. Work with the buyer for discounts

There are many emerging trends in procurement management. This question points to the need for large projects in different countries. When these projects involve procurement, it is important for the seller to work with the buyer prior to officially signing an agreement to take advantage of discounts through quantity purchases or other special considerations. [Executing]

PMI®, *PMBOK® Guide*, 2017, 463
PMI® *PMP Examination Content Outline*, 2015, Executing, 8, Task 1

164. a. Ensure your team knows the level of accepted risk appetite

    The effectiveness of risk management is related to project success. Management of risk keeps risk exposure within a defined range, and if implementing agreed-upon risk responses successfully, it can reduce negative variation, increase positive variation, and maximize achieving project objectives. However, to manage risks effectively, the team needs to know the risk appetite or level of risk exposure that is acceptable. Measurable risk thresholds can reflect the risk appetite of the organization and the project stakeholders. These thresholds show the degree of acceptable variation and should be stated explicitly and communicated to the project team. [Executing]

    PMI®, *PMBOK® Guide*, 2017, 397–398
    PMI® *PMP Examination Content Outline*, 2015, Executing, 8, Task 6

165. c. Hold a bidder conference

    A bidder conference is a tool and technique in Conduct Procurements. They also may be called contractor, vendor, or pre-bid conferences. The purpose is to have a meeting between the buyer and prospective sellers prior to proposal submittal. All prospective sellers then have a clear and common understanding, and no bidders receive preferential treatment. [Executing]

    PMI®, *PMBOK® Guide*, 2017, 487
    PMI® *PMP Examination Content Outline*, 2015, Executing, 8, Task 1

166. a. Have a meeting

    Meetings are a tool and technique in Plan Procurement Management. Research alone does not provide specific information to formulate a procurement strategy without influencing information exchange with potential sellers. However, the buyer organization can collaborate with potential bidders and may benefit. The sellers also can benefit to influence a mutually beneficial approach or product. Meetings then are used to determine strategy for managing and monitoring the procurement. [Planning]

    PMI®, *PMBOK® Guide*, 2017, 474
    PMI® *PMP Examination Content Outline*, 2015, Planning, 6, Task 7

167. c. Agreements for fast supply

Projects with high variability as in this question benefit from team structures that maximize focus and collaboration. Planning for physical and human resources then is less predictable. In such a situation, you need agreements for fast supply and team methods in order to control costs and to achieve the schedule objectives. [Planning]

PMI®, *PMBOK® Guide*, 2017, 312
PMI® *PMP Examination Content Outline,* 2015, Planning, 6, Task 5

168. d. Acquire generalized specialists

In agile, self-organizing teams are often used. In these teams, the team functions without centralized control. The project manager may not be called a project manager in these teams. Instead, the project manager provides the team with the environment and needed support and trusts the team to finish the project. In resource planning, successful self-organizing teams consist of generalized specialists rather than subject matter experts. These generalized specialists then continuously adapt to the changing environment. In acquiring these resources, they need to be people who can accept and even request constructive feedback. [Planning]

PMI®, *PMBOK® Guide*, 2017, 311
PMI® *PMP Examination Content Outline,* 2015, Planning, 6, Task 5

169. b. Use experts

Although all four answers are tools and techniques in the Plan Resource Management process, experts are the most appropriate answer. Expert judgment is a key tool and technique. It especially applies to this question as it notes expert judgment determines the preliminary effort level needed to meet project objectives. [Planning]

PMI®, *PMBOK® Guide*, 2017, 315
PMI® *PMP Examination Content Outline,* 2015, Planning, 6, Task 5

170. a. Preventive action

The Direct and Manage Project Work process, among other things, also includes change requests as an output. They can be submitted throughout the project, may be submitted by any stakeholder, and may be initiated from inside or outside the project. They also can be optional or legally or contractually mandated. They may require preventive action, the answer to this question, along with corrective action, defect repair, and updates. [Executing]

PMI®, *PMBOK® Guide*, 2017, 96
PMI® *PMP Examination Content Outline,* 2015, Executing, 7, Task 4

171. a. Review agreements

Agreements are an input to the Close Project or Phase. This situation had four contractors. The agreements require review since the requirements for formal procurement closure are defined in the contract terms and conditions and are in the procurement management plan. They are especially important to review in complex projects in which multiple contractors may be managed simultaneously or in sequence. [Closing]

PMI®, *PMBOK® Guide*, 2017, 125
PMI® *PMP Examination Content Outline,* 2015, Closing, 10, Task 3

172. b. Having an environment with an emphasis on transparent communications with your stakeholders

While the Manage Communications process covers numerous areas, in working in an agile environment, posting artifacts that are completed in a transparent fashion should be done. Holding regular stakeholder reviews also is recommended to promote communications with management and stakeholders. [Executing]

PMI®, *PMBOK® Guide*, 2017, 365
PMI® *PMP Examination Content Outline,* 2015, Executing, 8, Task 6

173. a. Adaptive life cycle

The adaptive life cycle is one that is known as agile or change driven and is set up to respond to high levels of change and ongoing stakeholder involvement. The detailed scope is defined and approved before the start of an iteration; they then can be referred to as agile, iterative, or incremental. Other life cycles are predictive or a hybrid. [Executing]

PMI®, *PMBOK® Guide*, 2017, 19
PMI® *PMP Examination Content Outline,* 2015, Executing, 8, Task 6

174. b. Gather data through focus groups

You are new to the organization and introducing formal project management to it represents a major culture change. Since most people tend to resist change, you need to determine the extent the people in the organization are ready for change and will accept it to enable you to determine how much tailoring you need to do. You can use focus groups as a data gathering technique to gauge the attitudes of the people involved and help you as you prepare your project management plan. With focus groups, you can bring stakeholders together to discuss your approach and help integrate different parts of the project management plan. Other data gathering techniques are brainstorming, checklists, and interviews; tools and techniques in the Develop Project Management Plan process. [Planning]

PMI®, *PMBOK® Guide*, 2017, 85
PMI® *PMP Examination Content Outline*, 2017, Planning, 7, Task 11

175. d. Correct identification may lead to project success

Academic research and analysis of high-profile projects that failed shows the importance of using a structured approach to identify, prioritize, and engage all stakeholders. One powerful but negative stakeholder may cause the project to fail. If the project manager can correctly identify and then later engage stakeholders in an appropriate way, it may lead to the difference between project success or failure. [Initiating]

PMI®, *PMBOK® Guide*, 2017, 504
PMI® *PMP Examination Content Outline*, 2015, Initiating, 5, Task 3

176. b. Select communications technology

In the Manage Communications process the choice of media is a key technique for communications management. It involves decisions concerning artifacts to meet the needs of the project. Examples are when to communicate in writing versus orally, when to prepare an informal memo rather than a formal report, when to use push or pull communications, and the choice of technology. Communications technology also is a tool and technique in Manage Communications. Factors to consider are the location of the team, the confidentiality of information to share, resources available to the team, and the organization's culture since it affects how meetings and discussions typically are conducted. [Executing]

PMI®, *PMBOK® Guide*, 2017, 381, 383
PMI® *PMP Examination Content Outline*, 2015, Executing, 8, Task 6

177. d. Storming

   During the storming stage, the team is addressing the work, technical decisions, and the project management approach. However, if team members are not collaborating and open to different ideas and perspectives, the environment becomes counterproductive. It is the second stage in the Tuckman team development [forming, storming, norming, performing, and adjourning] model. [Executing]

   PMI®, *PMBOK® Guide*, 2017, 338
   PMI® *PMP Examination Content Outline*, 2015, Executing, 7, Task 1

178. c. Prioritization

   All the answers are examples of data representation techniques used to categorize stakeholders. The key words in this project is that it is a large project, and 155 stakeholders were identified through stakeholder analysis. Prioritization is recommended when there is a large number of stakeholders, when the stakeholder community membership would change frequently, or when the relationship between the stakeholders and the project team or in the stakeholder community are complex. The latter two examples also characterize this situation. [Initiating]

   PMI®, *PMBOK® Guide*, 2017, 513
   PMI® *PMP Examination Content Outline*, 2015, Initiating, 5, Task 3

179. d. Emphasize flexible processes

   An emerging practice in project risk management is project resilience. These are risks that only are recognized after they occur. These emergent risks can be handled through project resilience by the answer to this question. Other ways to do so are a budget and schedule contingency for these risks, an empowered project team, frequent reviews of early warning signs, and input from stakeholders concerning areas where project scope or strategy can be adjusted if emergent risks occur. [Planning]

   PMI®, *PMBOK® Guide*, 2017, 399
   PMI® *PMP Examination Content Outline*, 2015, Planning, 6, Task 10

180. c. Organizational, stakeholder, and customer culture

This answer is an enterprise process asset to consider. People are more likely to want to share knowledge if they work in a culture that has a trusting relationship and a no-blame approach or shoot-the-messenger. Other factors are the value placed on learning and social behavioral norms. While the geographic locations of the team members are also an enterprise environmental factor, it does not fit this question, and the other two possible answers are project documents. [Executing]

PMI®, *PMBOK® Guide*, 2017, 101
PMI® *PMP Examination Content Outline*, 2015, Executing, 8, Task 6

181. b. Add risk management to the agreement

An emerging trend in procurement management is more advanced risk management. In Conduct Procurements contracts or agreements now may specify that risk management be performed as part of the contract for both the buyer and the seller. [Executing]

PMI®, *PMBOK® Guide*, 2017, 463
PMI® *PMP Examination Content Outline*, 2015, Executing, 8, Task 1

182. a. Observation

Observation/conversation is a key interpersonal and team skill and is a tool and technique in Manage Stakeholder Engagement. It is useful to stay in touch with the work and attitudes of the project team and stakeholders. Such an approach enables the project manager to have a better understanding of any concerns or issues stakeholders are having and to be able to then work to resolve them proactively. It also can be useful in avoiding potential conflicts among the team members or with any other stakeholders. [Executing]

PMI®, *PMBOK® Guide*, 2017, 527
PMI® *PMP Examination Content Outline*, 2015, Executing, 7, Task 7

183. d. Transference

Risk transfer is shifting some or all negative impact of a threat and the ownership of the response to the threat to a third party. It does not eliminate the threat posed by an adverse risk. It simply gives another party responsibility for it, and the other party is responsible if the risk occurs. It almost always involves paying a risk premium to the party taking on the risk. Transfer tools are diverse and include using insurance, performance bonds, warranties, and guarantees. Agreements also can be used to transfer ownership and liability for the risk to the third party. It may be prudent to transfer some work and its concurrent risk back to the buyer. [Planning]

PMI®, *PMBOK® Guide*, 2017, 443
PMI® *PMP Examination Content Outline*, 2015, Planning, 6, Task 10

184. b. 'As-built' plans

The key word in the answer is it is a construction program. The project manager working with contractors needs to review the procurement documentation. It includes collecting, indexing, and filing information on contract schedule, scope, quality, and cost performance. It also includes the contract change management system, payment records, and inspection reports and cataloging them for future reference. It also involves 'as-built' plans/drawings, or 'as-developed' documents, manuals, and other technical documentation. The purpose is to use this information for lessons learned and to evaluate contractors for future procurements. [Closing]

PMI®, *PMBOK® Guide*, 2017, 125
PMI® *PMP Examination Content Outline*, 2015, Closing, 10, Task 6

185. d. Review the business case

Even though this project was one with a lot of contracts in systems integration, as the project is closing, the business case is reviewed. It is reviewed since it documents the business need and the cost/benefit analysis that were used to justify the project. It is an input in the Close Project or Phase. [Closing]

PMI®, *PMBOK® Guide*, 2017, 125
PMI® *PMP Examination Content Outline*, 2015, Closing, 10, Task 1

186. c. Inherent systems complexity

There are two classes of non-event risks: variable risks and ambiguity risks. Inherent systems complexity is an example of an ambiguity risk. They are described as ones in which uncertainty exists about the future, or imperfect knowledge affects the project's ability to fulfill its objectives. Other examples in addition to the answer are elements of requirements or the technical solution or future developments in regulatory frameworks. [Planning]

PMI®, *PMBOK® Guide*, 2017, 398
PMI® *PMP Examination Content Outline*, 2015, Planning, 6, Task 10

187. d. Prompt lists

A prompt list is a tool and technique in Identify Risks as a predetermined list of risk categories. Its purpose is to aid the project team in generating ideas when using risk identification techniques. The categories in the lowest level of the RBS can be used as a prompt list. Some strategic frameworks such as political, economic, social technological, legal, environment or volatility, uncertainty, complexity, ambiguity may be used. [Planning]

PMI®, *PMBOK® Guide*, 2017, 416
PMI® *PMP Examination Content Outline*, 2015, Planning, 6, Task 10

188. b. The right message with the right content is delivered to the right audience

In the Monitor Communications process, the purpose is to ensure stakeholder information needs are met. The goal is an optimal information flow, which tracks to the answer to this question. Monitor Communications may require a variety of methods such as stakeholder satisfaction surveys, collecting lessons learned, observing the team, reviewing data from the issue log, and evaluating the stakeholder engagement matrix. [Monitoring and Controlling]

PMI®, *PMBOK® Guide*, 2017, 389
PMI® *PMP Examination Content Outline*, 2015, Monitoring and Controlling, 9, Task 1

189. b. Negotiation

Negotiation is the interpersonal and team skill used in Conduct Procurements. It is a discussion with the purpose to reach agreement. Procurement negotiation is needed to clarify the structure, rights, and obligations of the buyer and the seller. Its goal is mutual agreement before an agreement is signed. It concludes with a signed agreement that is executed the buyer and seller. [Executing]

PMI®, *PMBOK® Guide*, 2017, 488
PMI® *PMP Examination Content Outline*, 2015, Executing, 8, Task 1

190. a. Communicating whether the remaining reserve is adequate

    Burndown charts are used in reserve analysis, a data analysis tool and technique in Monitor Risks. Throughout the project negative and positive risks will occur with budget and schedule reserves needed to manage them. Reserve analysis then compares the amount of contingency reserve remaining to the amount of risk remaining to see if the reserve is adequate. Burndown charts show these data graphically. [Monitoring and Controlling]

    PMI®, *PMBOK® Guide*, 2017, 456
    PMI® *PMP Examination Content Outline*, 2015, Monitoring and Controlling, 9, Task 4

191. b. Quality and technical

    There are six commonly used source selection methods. All of the possible answers are ones to consider. In this situation, it is a high-risk project, and it is important that the selected seller have an outstanding quality-based/highest technical proposal score. You a using a RFP. The sellers are submitting a proposal with both technical and cost details. Using this method, the technical proposals are first ranked on the quality of the solution, and then the seller with the highest rating is invited to negotiate if the technical proposal is acceptable. Then, the financial proposal can be negotiated and evaluated. [Planning]

    PMI®, *PMBOK® Guide*, 2017, 474
    PMI® *PMP Examination Content Outline*, 2015, Planning, 6, Task 7

192. b. Categories of changes received

    Work performance information in Control Scope is an output of this process. It includes information on how the project is performing in relation to the scope baseline. It also can include the categories of the changes received, identified scope variances, and their causes, and how they impact schedule or cost and future scope performance. [Monitoring and Controlling]

    PMI®, *PMBOK® Guide*, 2017, 170
    PMI® *PMP Examination Content Outline*, 2015, Monitoring and Controlling, 9, Task 1

193. a. Organizational process asset

Learning reviews before, during, and after the project are an example of formal knowledge-sharing and information-sharing procedures, an organizational process asset. Identifying, capturing, and sharing lessons learned from this project and other projects complement these learning reviews. Organizational process assets are an input to the Manage Project Knowledge process. [Executing]

PMI®, *PMBOK® Guide*, 2017, 102
PMI® *PMP Examination Content Outline*, 2015, Executing, 8, Task 6

194. a. Create it early in the project

While the lessons-leaned register is an output of the Manage Project Knowledge process, it is best to create it early. In this process, it also is an input as it provides information on effective practices in knowledge management and is a project document. By creating it early, it can be used by other processes in the project; therefore, providing additional value. The people working on the project are responsible for capturing the lessons learned. Knowledge then can be documented by a variety of methods. [Executing]

PMI®, *PMBOK® Guide*, 2017, 101–104
PMI® *PMP Examination Content Outline*, 2015, Executing, 8, Task 6

195. a. Plan Procurement Management

Enterprise environmental factors, which include marketplace conditions and products, services, and results, are ones the team needs to be aware of as they develop plans for purchases and acquisitions. They are an input to the Plan Procurement Management process. [Planning]

PMI®, *PMBOK® Guide*, 2017, 470
PMI® *PMP Examination Content Outline*, 2015, Planning, 6, Task 7

196. d. Contributor

Stakeholder analysis is used as a data analysis technique in Identify Stakeholders. It results in a list of stakeholders, which includes among other things, their roles on the project. One role is that of a contributor. This is a person who makes a contribution to the project in terms of providing funds or other resources, including people, or provides support for the project in intangible ways. Intangible ways include advocacy in promoting project objectives or serving as a buffer between the project and the organization's power structure and its politics. [Initiating]

PMI®, *PMBOK® Guide*, 2017, 512
PMI® *PMP Examination Content Outline*, 2015, Initiating, 5, Task 3

197. d. Political awareness

While all of the answers are examples of interpersonal skills in the Manage Project Knowledge process, political awareness if the one that relies on helping the project manager plan communications based on the project and the organization's political environment. [Executing]

PMI®, *PMBOK® Guide*, 2017, 104
PMI® *PMP Examination Content Outline*, 2015, Executing, 8, Task 6

198. b. Bid documents

Bid documents are a procurement document, an input to Conduct Procurements. The procurement documents include the RFP, RFI, RFQ, or other documents sent to sellers for response. They need review before an agreement can be prepared and signed, [Executing]

PMI®, *PMBOK® Guide*, 2017, 485
PMI® *PMP Examination Content Outline*, 2015, Executing, 8, Task 1

199. b. The lessons-learned register becomes an organizational process asset

The lessons-learned register is an output of the Manage Project Knowledge process. However, at the end of the project or a phase, as in this situation, the information in the lessons-learned register then transfers to an organizational process asset. [Executing]

PMI®, *PMBOK® Guide*, 2017, 104
PMI® *PMP Examination Content Outline*, 2015, Executing, 8, Task 6

200. b. Where the stakeholder has the most influence

The stakeholder register has three main sections: identification information, assessment information, and stakeholder classification. This answer is in the assessment information section, which describes the major requirements, expectations, potential to influence the project outcome, and the phase of the project where the stakeholder has the greatest impact or influence. [Initiating]

PMI®, *PMBOK® Guide*, 2017, 515
PMI® *PMP Examination Content Outline*, 2015, Initiating, 5, Task 3

References

PMI® (*see* Project Management Institute).

Pritchard, Carl L., *Risk Management: Concepts and Guidance*. 5th ed. Boca Raton, FL: CRC Press., 2015.

Project Management Institute. PMI® Code of Ethics and Professional Conduct. www.pmi.org

_____. *A Guide to the Project Management Body of Knowledge, (PMBOK® Guide)*. 6th ed. Newtown Square, PA: Project Management Institute, 2017.

_____. *Organizational Project Management Maturity Model (OPM3®)*. 4th ed. Newtown Square, PA: Project Management Institute, 2018.

_____. *PMP Examination Content Outline*, June 2015. www.pmi.org

_____. *Practice Standard for Earned Value Management*, 2nd ed. Newtown Square, PA: Project Management Institute, 2011.

_____. *Project Management Professional (PMP)® Credential Handbook*, 2015. http://www.pmi.org/

_____. *The Standard for Portfolio Management—4th Edition*. Newtown Square, PA: Project Management Institute, 2017.

_____. *The Standard for Program Management—4th Edition*. Newtown Square, PA: Project Management Institute, 2017.

Rose, Kenneth H. *Project Quality Management: Why, What and How*. Second Edition. Boca Raton, FL: J. Ross Publishing, 2014.

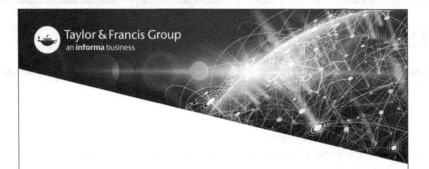